AF782415

The unknown gives birth to the known, yet remains unknown.

As a seed is the last of innumerable seeds, and contains the experience and the promise of numberless forests, so does the Unknown contain all that was, or could have been and all that shall or would be.

Sri Nisargadatta Maharaj

Even if today we can no longer agree with Kant in details, nevertheless the most general and fundamental idea of the Kantian epistemology retains its significance: to ascertain the 'a priori' intuitive mode of thought, and thereby to investigate the condition of the possibility of all knowledge. ....something is already given to us in advance in our faculty of representation: certain extra-logical concrete objects that exist intuitively as an immediate experience before all thought.

David Hilbert

Other Books of Interest

Algebra (Volume 1: Groups)
I.S. Luthar and I.B.S. Passi

Algebra (Volume 2: Rings)
I.S. Luthar and I.B.S. Passi

Algebra
V. Sahai and V. Bist

Analysis
E.H. Lieb and M. Loss

Analytical Geometry of Two Dimension
D. Chatterjee

Contemporary Abstract Algebra, 4/e
Joseph A. Gallian

A First Course in Mathematical Analysis
D. Somasundaram and B. Choudhary

Foundations of Complex Analysis
S. Ponnusamy

Functional Analysis: Selected Topics
P.K. Jain

Introduction to Linear Algebra
G. Schay

An Introduction to Measure and Integration
I.K. Rana

Nonlinear Functional Analysis
R. Akerkar

Nonlinearities in Complex Systems
S. Puri and S. Duttagupta

Sequence Spaces and Applications
P.K. Jain and E. Malkowsky

Topological Algebras
V.K. Balachandran

Forthcoming Books

Aspects of Combinatorics and Combinatorial Number Theory
S.D. Adhikari

A First Course in Lattice Theory and Boolean Algebra (With Applications)
Rita Malhotra

Mathematical Analysis and Applications
A.P. Dwivedi

Mathematics and Its Applications in Industry and Business
A.H. Siddiqui and K. Ahmad

Function Spaces and Applications
D.E. Edmunds et al

Sukumar Das Adhikari

An Introduction to **Commutative Algebra and Number Theory**

Narosa Publishing House
New Delhi Madras Bombay Calcutta
London

Sukumar Das Adhikari
The Mehta Research Institute of Mathematics and
Mathematical Physics
Allahabad, INDIA

Copyright © 1999 Narosa Publishing House

NAROSA PUBLISHING HOUSE

6 Community Centre, Panchsheel Park, New Delhi 110 017
22 Daryaganj, Prakash Deep, Delhi Medical Association Road, New Delhi 110 002
35–36 Greams Road, Thousand Lights, Madras 600 006
306 Shiv Centre, D.B.C. Sector 17, K.U. Bazar P.O., New Mumbai 400 705
2F–2G Shivam Chambers, 53 Syed Amir Ali Avenue, Calcutta 700 019
3 Henrietta Street, Covent Garden, London WC2E 8LU, UK

All rights reserved. No part of this publication may be reproduced, stored in a retrieval system, or transmitted in any form or by any means, electronically, mechanically, by photocopying, recording or otherwise, without the prior written permission of the publishers.

All export rights for this book vest exclusively with Narosa Publishing House. Unauthorised export is a violation of Copyright Law and is subject to legal action.

ISBN 81-7319-304-5

Published by N.K. Mehra for Narosa Publishing House, 6 Community Centre, Panchsheel Park, New Delhi 110 017 and printed at Rajkamal Electric Press, Delhi 110 033 (India).

TO MY TEACHER

Prof. R. Balasubramanian

Preface

This is an extended version of the notes which were prepared for the course of lectures given by the author at the **M**athematics **T**raining and **T**alent **S**earch programme (**MTTS**) in Pune (1996) and Chennai (1998).

The preparation of these notes started in an attempt to give a self-contained and elaborate version of the eleventh chapter (Factorization) of Michel Artin's book 'Algebra'. While the above mentioned book, to some extent, did give us a guideline regarding both the content and the presentation, the style of the end product is its own.

As for some of the main deviations from Artin, while discussing on algebraic numbers we have treated the subject with more generality, but not making the material too lengthy at the same time; there is a separate section on quadratic reciprocity, where we have supplied a complete proof of the theorem (an elementary proof has also been sketched in the exercises); a proof of the Lucas-Lehmer test is given in the appendix. On the whole, due to the personal inclination of the author, since its inception in the original MTTS lecture notes, the treatment had been kept close to the theory of numbers and that decided the choice of the title of the present book.

It is assumed that the reader has the basic knowledge of set theory. Eventhough the preliminaries on Groups, Rings and Fields are also assumed to be known, in Chapter 0, we provide a quick glance through them. In fact, upto Chapter 6, while covering some basic algebra, we establish many number theoretic results - sometimes as applications of algebraic results but not always. Apart from few exercise sets, there are many exercises all over the text. Some of these exercises are meant for recalling some results which would be used later, some others are meant for having a better understanding of a new result. Soloutions to selected exercises are given in the last chapter.

There are some remarks marked $*$, which are added without defining the terminologies involved and hence not supplying enough explanations as they would be beyond the scope of this book. However, it was felt that such remarks may serve as stimulants to further studies. Theorems, exercises and remarks are numbered in separate sequences. Definitions, notes and examples are not numbered. While numbering a corollary of a particular theorem, it is done by puttiung the number of the related theorem followed by an additional serial number.

The bibliography gives informations regarding sources for further knowledge on the subject.

I would like to thank : Professor S. Kumaresan (Mumbai) for inviting me to deliver the lectures in the MTTS programme; Professors Yong-Gao Chen and Chungang Ji (Nanjing) for providing me with several historical informations about the Chinese Remainder Theorem as well as the Chinese poem which appears in the text; Professors D. Prasad, Dinesh Thakur (Arizona), S.A. Katre (Pune) and Ramji Lal (Allahabad) for going through parts of the manuscript; Mr. Anirban Mukhopdhyay, Mr. R. Thangadurai, Mr. Nilabh Sanat, Dr. Jaya Iyer and Dr. Suresh Nayak for helping me in proof-reading and offering some suggestions at the time of the preparation of the manuscript. I am also thankful to my friend and colleague Dr. B. Ramakrishnan for constant encouragement.

Sukumar Das Adhikari

Notations and terminologies

Throughout the text, the symbols $\mathbf{N}, \mathbf{Z}, \mathbf{Q}, \mathbf{R}$ and $\mathbf{C}$ will denote respectively the set of non-negative integers, the set of integers, the set of rationals, the set of real numbers and the set of complex numbers.

The word *ring* will always mean a commutative ring with identity.

A function $f : A \to B$ is *injective* if $f(a_1) = f(a_2)$ implies $a_1 = a_2$; f is said to be *surjective* if for any $b \in B$ there is $a \in A$ such that $f(a) = b$.

For two sets A and B, the notation $A < B$ will be used to denote that A is strictly contained in B. For a finite set S, $|S|$ will denote the number of elements of S.

For a ring R, the symbol R^* will denote its group of units.

For any prime power q, $\mathbf{F}_q$ will denote the finite field with q elements. For a real x, by $[x]$ we denote its integral part.

The symbol $\square$ is used to denote the end of a proof.

Contents

Chapter 0

Preliminaries: Groups, rings and fields

0.1 Groups

Definition. Let G be a non-empty set. A *law of composition* on G is a mapping $f : G \times G \to G$. One uses notations like $a.b$, $a + b$ or a simple juxtaposition ab for $f((a, b))$.

Definition. Let G be a non-empty set together with a law of composition $f : G \times G \to G$. We write ab for $f((a, b))$. Then the pair (G, f) is called a *group* if the following conditions are satisfied:

i) $a(bc) = (ab)c$ for all a, b, c in G. This fact is expressed by saying that the law of composition is *associative*.

ii) There is an element e in G such that $ea = ae = a$ for all $a \in G$. Clearly the element e is unique; it is called the *identity element* of G.

iii) For each $a \in G$ there is an element a^{-1} in G such that $aa^{-1} = a^{-1}a = e$. By the associative law, for any $a \in G$ such an element a^{-1} is unique and it is called the *inverse* of a.

Definition. If the pair (G, f) in the above definition satisfies the following additional condition then we call it an *abelian group*:

iv) For all a, b in G, $ab = ba$. This fact is expressed by saying that the law of composition is *commutative*.

Examples. The set $\mathbf{Z}$ of integers with usual addition as law of composition is an abelian group. The set of non-zero rationals $\mathbf{Q}^*$ is also an abelian group with multiplication as law of composition.

Note. From now on for a group (G, f) we shall often write G, assuming that the map f involved is known. Because of the associativity of the law of composition one can use the notation abc for $a(bc) = (ab)c$. In general, in the case of a string of n elements, one uses notations like $a_1a_2 \cdots a_n$.

If $a_i = a$ for all $i = 1, 2, \cdots, n$, one writes a^n to denote $\underbrace{a \cdots a}_{n \text{ times}}$.

For a positive integer n, a^{-n} is defined to be $\underbrace{a^{-1} \cdots a^{-1}}_{n \text{ times}}$.

$a^0 = a^{1-1}$ is the identity element e.

If one chooses the additive notation instead of juxtaposition, then one writes na for the element $\underbrace{a + \cdots + a}_{n \text{ times}}$.

Definition. Let H be a subset of a group G. Then H is called a subgroup of G if the following conditions are satisfied:

i) For all $a, b \in H$, $ab \in H$.

ii) The identity element $e \in H$.

iii) If $a \in H$, then $a^{-1} \in H$.

Thus, if H is a subgroup of G, then it is a group under the law of composition induced from G.

Example. For any integer a, the subset $a\mathbf{Z} \overset{\text{def}}{=} \{ar | r \in \mathbf{Z}\}$ is a subgroup of $(\mathbf{Z}, +)$.

Definition. Let G be a group and a an element of G. Let $H = \{a^n | n \in \mathbf{Z}\}$. Then it is easy to see that H is a subgroup of G. In fact, it is the smallest subgroup of G containing a, that is, any subgroup of G containing the element a will contain H. H is called the *cyclic subgroup* of G *generated by a*.

A group G is called *cyclic* if there is an element $a \in G$ such that the cyclic subgroup generated by a is the whole of G. We say that the element *a generates* (or a is a *generator* of) the cyclic group G.

The *order of a group* G is the number of elements of G and *the order of an element* a is the order of the cyclic subgroup generated by it. If the order of a group is finite, it is called a *finite group*. Otherwise, it is called *infinite*.

Example. The group $\mathbf{Z}$ under addition is an infinite cyclic group. Both the elements 1 and -1 are generators of $\mathbf{Z}$.

Definition. A *homomorphism* of a group G into a group G' is a map $\phi : G \to G'$ such that $\phi(ab) = (\phi(a))(\phi(b))$ for every $a, b \in G$. A homomorphism $\phi : G \to G'$ of a group G into a group G' is called an *isomorphism* if it is both injective and surjective. If there exists an isomorphism $\phi : G \to G'$, then the groups G and G' are said to be *isomorphic*.

Remark 0.1. Let G and G' be two groups with e and e' as the respective identity elements of the groups. If $f : G \to G'$ is a homomorphism, then $f(e) = f(ee) = f(e)f(e)$ and multiplying both sides by $(f(e))^{-1}$, we have $f(e) = e'$. Also, for any element $a \in G$, $f(aa^{-1}) = f(e) = e'$ and hence $f(a^{-1}) = (f(a))^{-1}$.

Examples. Let $f : \mathbf{Z} \to \mathbf{Z}$ be defined by $f(n) = 2n$ for $n \in \mathbf{Z}$. Then f is a homomorphism. In general, given a group G, if b is an element of G, then the map $\phi : \mathbf{Z} \to G$ defined by $\phi(n) = b^n$ is a homomorphism.

Let $\mathbf{Q}^*$ be the group of non-zero rationals under multiplication. Then, the map $f : \mathbf{Q}^* \to \mathbf{Q}^*$ defined by $f(\alpha) = \alpha^{-1}$ for all $\alpha \in \mathbf{Q}^*$ is an isomorphism.

Definition. Let G and G' be groups and $f : G \to G'$ a homomorphism. Then the *kernel* of f is defined to be the set of all elements $x \in G$ such that $f(a) = e'$ where e' is the identity element of G'. From Remark 0.1 above, the identity element e of G belongs to the kernel of f. It is easy to see that the kernel of f is a subgroup of G and f is injective if and only if it is equal to $\{e\}$.

Definition. An *equivalence relation* on a set S is a subset R of $S \times S$ satisfying the following three conditions:

i) Reflexive: $(a, a) \in R$ for every element $a \in S$.

ii) Symmetric: If $(a, b) \in R$, then $(b, a) \in R$.

iii) Transitive: If (a, b) and (b, c) are both in R, then (a, c) is also in R.

If $(a, b) \in R$, we say that a is equivalent to b and write $a \sim b$.

It is clear that given an equivalence relation R on a set S, we have a partition of S into disjoint subsets where each subset will consist of elements which are equivalent to each other. Thus if a belongs to a subset, then any b in that subset will satisfy $a \sim b$. This subset consisting of all elements equivalent to a is called the *equivalence class* of a and is denoted by C_a. If $a \sim b$, then $C_a = C_b$. For a given equivalence relation R on a set S, the set of all equivalence classes is called the *quotient* of S by R and is denoted by S/R or $S/\sim$.

In the remaining part of this section, we shall consider abelian groups only.

Definition. Let G be an abelian group and H a subgroup of G. Let R be the subset of $G \times G$ consisting of those elements (a, b) such that $ab^{-1} \in H$. It is easy to see that R is an equivalence relation on G. The equivalence classes here are called *cosets of G modulo H* and the quotient is said to be the *quotient of* G by H and is denoted by G/H. One can check that the quivalence class of a is $aH \stackrel{\text{def}}{=} \{ah | h \in H\}$. Defining multiplication in G/H by $(aH)(bH) = (ab)H$, G/H becomes a group called the *quotient group* of G by H.

Remark 0.2. The number of cosets of a subgroup H is called the *index* of H in G and it is denoted by $[G : H]$. Now, an equivalence relation given on a set gives a decomposition of the set into a union of mutually disjoint equivalence classes. Therefore, if G is a finite group, then observing that each coset of H has the same number of elements as that of H, we have $|G| = [G : H]|H|$, which implies that the order of H divides that of G. (The statement that the order of a subgroup of a finite group divides the order of the group is true even if the group is not abelian and this result is known as *Lagrange's theorem* in the literature.)

If we define $\phi : G \to G/H$ by $\phi(a) = aH$, then ϕ is a surjective homomorphism with kernel H.

Now, let G and G' be groups and f is a homomorphism from G into G'. Let H be the kernel of f. Then we consider the homomorphism $\overline{f}$ from G/H to the image $f(G)$ of f defined by $\overline{f}(aH) = f(a)$. It is an isomorphism of G/H onto $f(G)$. This is known as *the first isomorphism theorem.*

0.2 Rings and fields

Definition. A *commutative ring with identity* R is a set with two laws of compositions, called *addition* (denoted by $+$) and *multiplication* (denoted by juxtaposition), satisfying the following conditions:

i) R is an abelian group with respect to addition. The identity element of $(R, +)$ is called the *zero element* of R and is denoted by 0. For $x \in R$, the additive inverse is denoted by $-x$.

ii) Mutiplication is associative and it distributes over addition. That is, for any x, y, z in R, $(xy)z = x(yz)$ and $x(y + z) = xy + xz$, $(y + z)x = yx + zx$.

iii) There is $1 \in R$ such that $x1 = 1x = x$. This element 1 (which is clearly unique) is called the *identity element* of R.

iv) For all $x, y \in R$, $xy = yx$. This condition is called *commutativity.*

Remark 0.3. As mentioned earlier, we shall be using the word ring to mean a commutative ring with identity, that is, for one satisfying the conditions iii) and iv) in addition to the standard conditions i) and ii). We also note that because of our commutative condition iv), either of the distributive laws in ii) implies the other.

Remark 0.4. The possibility of 1 becoming equal to 0 is not excluded. If that happens, for any $x \in R$, we have $x = x1 = x0 = 0$. Thus, in that case R contains the single element 0 and is called the *zero ring.*

Examples. The set of integers $\mathbf{Z}$ is a ring under usual addition and multiplication. The same is true for the set of rationals $\mathbf{Q}$, the set of reals $\mathbf{R}$ and the set of complex numbers $\mathbf{C}$.

Definition. A *subring* S of a ring R is a subgroup of $(R, +)$ such that $1 \in S$ and $xy \in S$ for all $x, y \in S$.

Definition. If R and S are two rings, then a map $f : R \to S$ is called a *ring homomorphism* or simply a homomorphism (when the context is clear) if the following conditions are satisfied:

i) f(x+y) = f(x)+ f(y)

ii) f(xy) = f(x)f(y)

iii) f(1) = 1.

Remark 0.5. Here we have used the same symbol 1 to denote the identity element of R as well as that of S. It should be clear from the context which one is being referred to in a particular place. Similarly, in both the rings R and S, the symbol $+$ and juxtaposition are used respectively for addition and multiplication.

Definition. As in the case of groups, a ring homomorphism $f : R \to S$ is called an *isomorphism* between R and S if it is both injective and surjective. If there is an isomorphism from a ring R onto a ring S, then these two rings are said to be *isomorphic.*

Definition. An *ideal* I of a ring R is a subset of R which is a subgroup of $(R, +)$ such that if $a \in I$ and $r \in R$, then $ra \in I$.

For any $x \in R$, the set $Rx \stackrel{\text{def}}{=} \{rx | r \in R\}$ is clearly an ideal. It is called the *principal ideal* generated by x. Clearly, it is the smallest ideal containing x. When the ring R remains fixed during a certain discussion, we shall sometimes write (x) in place of Rx. In general, for any subset D of R, the smallest ideal containing D is called the *ideal generated by* D and it is the set of all finite sums of the form $r_1d_1 + r_2d_2 + \cdots + r_md_m$, $r_i \in R$ and $d_i \in D$ for $i = 1, 2 \cdots, m$ as m varies over positive integers.

Examples. For a ring R, the singleton set $\{0\}$ and the whole ring R are obviously ideals. Both of them are principal ideals generated respectively by 0 and 1.

Definition. An ideal of a ring R other than the ideals $\{0\}$ and R, mentioned above, is called a *proper ideal.*

Definition. If A is an ideal of a ring R, then the quotient group R/A inherits a multiplication from that of R (that is, (x +A)(y+A) = xy +A) which makes it a ring. This ring R/A is called the *quotient ring* of R by the ideal A. The map from R to R/A, sending x to $x + A$ can be easily seen to be a surjective homomorphism onto R/A.

Remark 0.6. Let $f : R \to S$ be a ring homomorphism. The set of all elements of R which are mapped to 0 is called the *kernel* of f and is written ker (f). It is easy to see that ker(f) is an ideal of R. The homomorphism f will induce a ring isomorphism from the quotient ring $R/$ker (f) to the image $f(R)$ of f.

Definition. An element x of a ring R is called a *zero-divisor* if there is $y \in R$, $y \neq 0$ such that $xy = 0$.

A non-zero ring R (that is, with $1 \neq 0$) is called an *integral domain* if it contains no zero-divisors other than 0.

A *unit* in R is an element u such that $uv = 1$ for some $v \in R$. The units of R form an abelian group under multiplication. It can also be observed easily that an element u of a ring R is a unit if and only if the principal ideal generated by u is R.

A *field* is an integral domain in which every element other than 0 is a unit.

A subring K of a field F is called a *subfield* of F, if K is a field under the operations induced from F. If K is a subfield of F, then F is called an *extension field* of K.

Definition. For a ring R with identity element 1, by an *R-module* one means an abelian gooup $(V, +)$ together with a scalar multiplication $R \times V \to V$ such that for all $r, s \in R$ and $v, v' \in V$

a) $1v = v$

b) $(rs)v = r(sv)$

c) $(r + s)v = rv + sv$

d) $r(v + v') = rv + rv'$.

Examples. **Z** is an integral domain. The integral domains **Q**, **R** and **C** are examples of fields. The fields **Q** and **R** are subfields of **C**. Abelian groups are nothing but **Z**-modules. A ring R is a module over itself and its ideals are precisely the R-submodules of R. If F is a field, an F-module is just an F-vector space. We assume that the reader is familier with vector spaces. We shall take up the study of modules in Chapter 8.

Exercise 0.1. The only ideals of a field F are $\{0\}$ and F. Conversely, if a ring R has no proper ideal, then it is a field.

Exercise 0.2. Show that a finite integral domain is a field.

Definition. An ideal P of a ring R is called a *prime ideal* if $P \neq R$ and for $x, y \in R$, if the product $xy \in P$, then $x \in P$ or $y \in P$.

An ideal M of a ring R is called a *maximal ideal* if $M \neq R$ and there is no ideal A satisfying $M < A < R$.

Remark 0.7. It is not difficult to observe that an ideal P of R is prime if and only if R/P is an integral domain. Similarly, an ideal M is maximal if and only if R/M is a field. Clearly, a maximal ideal is prime. The converse is not true in general.

Exercise 0.3 Show that every non-zero ring R has at least one maximal ideal.

We end this section after sketching a proof of the following important theorem.

Theorem 0.1. Given any integral domain D, there is a field F containing an isomorphic image of D as a subring.

Proof: Let S be the set $\{(a,b)|a,b \in D, b \neq 0\}$. For elements of S, we define $(a,b) \sim (c,d)$ if $ad = bc$. It is clearly an equivalence relation as D does not contain any zero-divisor. Given an element (a,b) of S, we denote the equivalence class of (a,b) by a/b. We then observe that the addition and multiplication defined by $a/b + c/d = (ad+bc)/bd, (a/b)(c/d) = (ac/bd)$ in the quotient $S/\sim$ are well-defined and makes it a ring.

We denote this ring by F. $1/1$ is the identity element of F. Now, if a/b is a non-zero element of F, then $a \neq 0$ and b/a is an inverse for it. Thus, F is a field. Observing that the map $i : D \to F$ defined by $i(a) = a/1$, is an injective homomorphism of D into F, we are through. □

Exercise 0.4. Given an integral domain D, let F be constructed as in the above proof. Show that given any injective homomorphism $f : D \to K$, where K is a field, f can be extended to an isomorphism of F into K in a unique way.

Definition. The field F in the proof of Theorem 0.1 is called the *quotient field* of D. Exercise 0.4 shows that F is the 'smallest' field containing D.

Chapter 1

The integers

In this chapter, we dwell briefly on the divisibility theory and some congruence results related to the set of integers **Z**. The axiomatic construction of the set of integers, as well as some basic properties of the set of integers, like the principle of mathematical induction and the fact that a non-empty set of positive integers contains a smallest element, will be assumed to be known. In fact, we had several occasions to refer to the set **Z** in the last chapter and we observed that it is a ring under usual addition and multiplication.

1.1 Divisibility

Definition. Let a be a non-zero integer. An integer b is said to be *divisible* by a, if there is an integer x such that $b = ax$ and this is expressed by writing $a \mid b$. We also express the situation by saying that a *divides* b, a is a *divisor* of b or b is a *multiple* of a.

In case b is not divisible by a, we write $a \nmid b$. If $a \mid b$ and $1 < |a| < |b|$, then a is called a *proper divisor* of b. The notation $a^k \parallel b$ will be used to indicate that $a^k \mid b$ and $a^{k+1} \nmid b$.

Theorem 1.1. (The division algorithm) Given two integers a and b, with $a > 0$, there exist unique integers q and r such that

$$b = qa + r, \ 0 \le r < a.$$

(One says that q is the *quotient* and r is the *remainder* obtained when b is divided by a. If $a \nmid b$, then r satisfies the stronger inequalities $0 < r < a$).

Proof: Since $a \geq 1$, we have $b + |b|a \geq b + |b| \geq 0$. Therefore, the set $S = \{b - xa | x$ is an integer and $b - xa \geq 0\}$ is nonempty and by the well ordering principle, will contain a smallest integer, say r. Now, r being an element of S will be of the form $b - qa$. If $r \geq a$, then $r - a = b - (q+1)a$ being also an element of S will contradict the fact that r is the smallest element of S. Hence $r < a$.

We now sketch how to observe the uniqueness. Suppose that there is another such pair q', r'. That will imply $|r - r'| = a|q - q'|$ as well as $|r - r'| < a$. From these two inequalities it follows that $|q - q'| < 1$. Since q and q' are integers, $q = q'$ and hence $r = r'$. □

Note. We assumed $a > 0$. In general, the division algorithm says that for integers a, b with $a \neq 0$, there exist q, r such that $b = aq + r, 0 \leq r < |a|$.

Exercise 1.1.

(*a*) Determine the subgroups of the additive group $(\mathbf{Z}, +)$ of integers.

(*b*) Show that a cyclic group is either isomorphic to the additive group of integers $\mathbf{Z}$ or to the quotient group $\mathbf{Z}/m\mathbf{Z}$ for some integer $m > 0$.

(*c*) Prove that every ideal in the ring $\mathbf{Z}$ of integers is a principal ideal.

Note. Recall that an ideal is called *principal* if it is generated by a single element. An integral domain, all of whose ideals are principal is called a *principal ideal domain* (PID). Therefore, part (*c*) of the above exercise says that $\mathbf{Z}$ is a PID.

Theorem 1.2. Let $a, b \in \mathbf{Z}$, not both zero. Then there is a common divisor d of a and b of the form $d = ar + bs$, for some integers r and s such that if an integer e divides a and b, it also divides d.

Proof: We consider the set

$$S = \{ax + by | ax + by > 0, x, y \in \mathbf{Z}\}.$$

Since a and b are not both zero, it is easy to see that S is nonempty. Therefore, S will contain a smallest element d. Since $d \in S$, $d = ar + bs$ for some $r, s \in \mathbf{Z}$.

Let m be any element of S. By the division algorithm, $m = qd + c$ where $0 \leq c < d$. If $m = ar_1 + bs_1$, $r_1, s_1 \in \mathbf{Z}$, then $c = m - qd = a(r_1 - qr) + b(s_1 - qs)$ which implies that $c \in S$. Therefore, the choice of d forces $c = 0$. Hence $m = qd$. In other words, d divides m.

In the above paragraph, we have proved that given any element m of S, d divides m. But, the integers a ($= a.1 + b.0$) and b (=a.0+b.1) belong to S. Hence d divides a and b.

If an integer e divides a and b, then e divides $d = ar + bs$ and the proof is complete. □

Definition. For obvious reasons, the integer d of Theorem 1.2 is called a *greatest common divisor (gcd)* of a and b and is denoted by (a, b).

If $(a, b) = 1$, then a and b are said to be *relatively prime.*

For an integer $n \geq 1$, the *Euler's totient ($\phi(n)$)* $\phi(n)$ is defined to be the number of positive integers not exceeding n which are relatively prime to n.

An integer $p > 1$ will be called a *prime number* (or simply a prime) if the only positive divisors of p are 1 and p. Therefore, for a prime p, $\phi(p) = p - 1$.

Given a positive integer n, two integers a and b are said to be *congruent modulo n*, written

$$a \equiv b \pmod{n},$$

if $n \mid (b - a)$, that is, $b - a = nk$ for some integer k, which is equivalent to saying that $b-a$ belongs to the ideal $n\mathbf{Z}$. As we had already observed in the last section, this is an equivalence relation. The equivalence classes (cosets) here are called *congruence classes* modulo n or *residue classes* modulo n. The congruence class of an integer a will be denoted by the symbol $\overline{a}$.

A set of n representatives, one from each residue classes modulo n, is called a *complete residue system* modulo n.

By a *reduced residue system* modulo n, one means a set of $\phi(n)$ integers, from distict residue classes, such that each one is relatively prime to n.

Remark 1.1. It can be easily seen that the index $[\mathbf{Z} : n\mathbf{Z}]$ of the subgroup $n\mathbf{Z}$ in $\mathbf{Z}$ is n. We already know that the map $\mathbf{Z} \longrightarrow \mathbf{Z}/n\mathbf{Z}$ sending an integer a to its congruence class $\overline{a}$ is compatible with addition and multiplication.

Theorem 1.3. (Euclid's lemma) If $a \mid bc$ and $(a, b) = 1$, then $a \mid c$.

Proof: Since $(a, b) = 1$, one can write $1 = ax + by$ for integers x and y. Therefore,

$$\begin{aligned} c &= acx + bcy \\ &= acx + ary \\ &= a(cx + ry) \end{aligned}$$

where $bc = ar$ as $a \mid bc$. □

Theorem 1.4. If a prime p divides ab, then $p \mid a$ or $p \mid b$. More generally, if a prime p divides a product $a_1 a_2 \cdots a_n$, then p divides at least one of the factors.

Proof: Assume $p \mid ab$ and $p \nmid a$. Now, $(p, a) = 1$. Therefore by Euclid's lemma, $p|b$. The second part follows easily by induction. □

1.2 The fundamental theorem of arithmetic

Theorem 1.5. (Fundamental theorem of arithmetic) Every integer $a \neq 0$ can be written as a product

$$a = cp_1 p_2 \cdots p_k,$$

where $c = \pm 1$, $p_i > 0$ are prime integers (not necessarily distinct) and $k \geq 0$. The expression is unique upto the ordering of the prime factors.

Proof: First we prove that a prime factorization exists. It is enough to consider the case $a > 1$. We proceed by induction. For $a = 2$, the result is true. By induction hypothesis, factorization exists for all positive integers $b < a$. If a is a prime, then there is nothing to prove. If a is not a prime, then $a = bb'$ where $b, b' > 1$. Now b and b' being strictly smaller than a, by induction hypothesis, they can be factored into primes. Setting these factorizations side by side gives a factorization of a.

It remains to show that the factorization is unique.

Suppose $\pm p_1 p_2 \cdots p_n = \pm q_1 q_2 \cdots q_n$ be two such factorizations. Applying Theorem 1.4 with $p = p_1$, we have $p_1 \mid q_i$ for some $i, 1 \leq i \leq m$. Since q_i is also prime, $p_1 = q_i$. Now we cancel p_1 and proceed by induction. □

***Remark 1.2.** The *Riemann zeta function* $\zeta(s)$ $(s = \sigma + it)$, which for $\sigma > 1$ is defined as $\sum_{n=1}^{\infty} \frac{1}{n^s}$ and is continued meromorphically over the complex plane, conceals in its bosom deep informations regarding the multiplicative nature of integers because of the identity

$$\sum_{n=1}^{\infty} \frac{1}{n^s} = \prod_p \left(1 - \frac{1}{p^s}\right)^{-1}, \ \sigma > 1.$$

This identity is an anlytic equivalent of Theorem 1.5 above.

Theorem 1.6. (Euclid) There are infinitely many prime numbers.

Proof: Suppose $p_1, p_2, \cdots, p_n$ are finitely many prime numbers with $1 < p_1 < p_2 < \cdots < p_n$. Let $N = 1 + p_1 p_2 \cdots p_n$. By fundamental theorem of arithmetic, there exists a prime number $p > 1$ which divides N. If $p = p_i$ for some $i, 1 \le i \le n$ it will imply that $p_i \mid 1$, which is impossible.

Thus, we have proved that given a finite number of primes, there is always a prime different from all of them. This proves the theorem. □

Exercise 1.2. Imitate the proof of the above theorem to prove that there are infinitely many primes of the form $4n + 3$.

***Remark 1.3.** Dirichlet's theorem on primes in arithmetic progressions says that for positive integers m and a, the arithmetic progression $mn + a$, $n = 1, 2, \cdots$ contains infinitely many primes if $(a, m) = 1$. In Exercises 1.6, 7.3 and D.4 (iv), we shall see some of the instances where, backed by some additional number theoretic knowledge, an Euclid type of argument yields the result for some special arithmetic progressions. However, Euclid's proof can not be pushed to yield the most general result of Dirichlet.

Dirichlet's original proof, which is analytic, had its origin in Euler's proof of the infinitude of primes. Euler's proof runs as follows:

For $x \ge 2$ and integer m such that $2^m \ge x$,

$$\prod_{p \le x} \left(1 - \frac{1}{p}\right)^{-1} > \prod_{p \le x} \left(1 + \frac{1}{p} + \cdots + \frac{1}{p^m}\right) \ge \sum_{n=1}^{[x]} \frac{1}{n},$$

from which it follows that $\prod_{p \le x} \left(1 - \frac{1}{p}\right)^{-1}$ diverges as $x \to \infty$.

Exercise 1.3.

(a) If m, m' are positive integers such that $(m, m') = 1$ and a, a' runs through complete sets of residues modulo m and m' respectively, then show that $a'm + am'$ runs through a complete set of residues mod mm'.

(b) If m, m' are as in part (a) above, show that $\phi(mm') = \phi(m)\phi(m')$, where ϕ is Euler's totient. One expresses this fact by saying that the function ϕ is *multiplicative.*

(c) Let $n > 1$ be an integer. If $n = p_1^{a_1} p_2^{a_2} \cdots p_k^{a_k}$, where p_i's are distinct primes, then

$$\phi(n) = n \prod_{i=1}^{k} \left(1 - \frac{1}{p_k}\right).$$

Exercise 1.4. If $(a, m) = 1$, then the linear congruence $ax \equiv b \pmod{m}$ has exactly one solution (modulo m).

Exercise 1.5. (Euler-Fermat) Let $(a, m) = 1$. Then $a^{\phi(m)} \equiv 1 \pmod{m}$ where ϕ is the Euler's totient.

Exercise 1.6. Let p be an odd prime. If there is a natural number n such that $n^2 + 1$ is divisible by p, then p is of the form $4m + 1$. (The converse is also true. See Theorem 3.1 (iii).) Deduce that there are infinitely many primes of the form $4m + 1$.

Exercise 1.7. Does the equation $x^2 + 1 = 3y^5$ have solutions in integers?

Remark 1.4. Equations to be solved in integers are called *Diophantine equations* after Diophantus (AD 250) of Alexandria. In this terminology, the above exercise can be rephrased by asking whether the Diophantine equation $x^2 + 1 = 3y^5$ has solutions.

1.3 The Chinese Remainder Theorem

The Chinese remainder theorem, which we are going to discuss now, originates from ancient publications. Interested readers may look into a recent book of Ding, Pei and Salomaa [DPS 1996] which gives diverse applications of this important theorem. Around the first century A.D., the Chinese remainder theorem was proposed by Sun Zi through a particular problem. Later, in 1593, Cheng Dawei explained Sun Zi's solution with a song. We give below the original Chinese song followed by an English translation.

"San ren tong xing qi shi xi,
wu shu mei hua nian yi zhi,
qi zhi tuan yuan zheng ban yue,
chu bai ling wu bian de zhi."

("Three people walking together, 'tis rare that one is seventy.
Five cherry blossom trees, twenty one branches bearing flowers.
Seven disciples reunite for the half moon.
Take away one hundred and five and you shall know".)

From the proof of the Chinese remainder theorem, we shall see that this riddle like song actually indicates the simultaneous solution in x of the following set of congruences:

$$\begin{aligned} x &\equiv b_1 \pmod 3 \\ x &\equiv b_2 \pmod 5 \\ x &\equiv b_3 \pmod 7. \end{aligned}$$

Theorem 1.7. (Chinese remainder theorem) Suppose $m_1, m_2, \cdots, m_r$ are positive integers, pairwise relatively prime, that is, $(m_i, m_k) = 1$ if $i \neq k$. Then, given arbitrary integers $b_1, b_2, \cdots, b_r$, the system of congruences

$$\begin{aligned} x &\equiv b_1 \pmod{m_1} \\ x &\equiv b_2 \pmod{m_2} \\ \cdots & \quad \cdots \\ x &\equiv b_r \pmod{m_r} \end{aligned}$$

has exactly one solution modulo the product $m_1 m_2 \cdots m_r$.

Proof: Let $M = m_1 m_2 \cdots m_r$. Let $M_k = M/m_k$. Now, $(M_k, m_k) = 1$ and hence M_k has a unique reciprocal M_k' modulo m_k.

Let $x = b_1 M_1 M_1' + b_2 M_2 M_2' + \cdots + b_r M_r M_r'$. Since $M_i \equiv 0 \pmod{m_k}$, for $i \neq k$, we have $x \equiv b_k M_k M_k' \equiv b_k \pmod{m_k}$. Hence x satisfies every congruence in the system. If x and y are two solutions of the system, $x \equiv y \pmod M$. □

Exercise 1.8. Solve the following system of simultaneous congruences:

$$\begin{aligned} x &\equiv 2 \pmod 3 \\ x &\equiv 3 \pmod 5 \\ x &\equiv 2 \pmod 7). \end{aligned}$$

Remark 1.5. At this point it is advisable to read the Chinese song again. While working out the above exercise along the line of the proof of Theorem 1.7, one has (with the notations as in the theorem) $r = 3$ and $M = 105$. The song supplies the numbers $M_i M_i'$, for $i = 1, 2$ and 3. They are respectively $35 \times 2 = 70, 21 \times 1 = 21$ and $15 \times 1 = 15$. (Here one had to understand that half moon refers to half of the number of days in a lunar month, that is 15.) They have to be multiplied respectively by the residues $2, 3$ and 2 (b_i's of the theorem) corresponding respectively to the primes $3, 5$ and 7. After adding the numbers thus obtained, one has to take away (that is, to subtract) multiples of $M = 105$ to get the smallest solution.

Exercise 1.9. Let $F(x)$ be a polynomial (for the definition of a polynomial, see Chapter 2, if needed) with integer coefficients. Let $m_1, m_2, \cdots, m_r$ be positive integers, relatively prime in pairs. Let $m = m_1 m_2 \cdots m_r$. Prove that the congruence $F(x) \equiv 0 \pmod m$ has a solution if, and only if, each of the congruences $F(x) \equiv 0 \pmod{m_i}$, $i = 1, 2, \cdots, r$ has a solution.

Exercise 1.10.

(a) If p is an odd prime, show that there are integers a and b such that $a^2 + b^2 + 1 \equiv 0 \pmod p$.

(b) Show that the above result holds even if p is replaced by p^r where p is an odd ptime and $r \geq 1$ is any integer.

(c) Hence deduce that given any odd positive integer m, there are integers a and b such that $a^2 + b^2 + 1 \equiv 0 \pmod m$.

Chapter 2

Polynomial rings

In this chapter, we start from the formal definition of polynomials with coefficients in a ring. Making a ring out of the set of polynomials with coefficiants in a ring, we get a new supply of rings which will turn out to be very important. We also derive some fundamental properties of the ring of polynomials over a field, which will play an important role in the study of field extensions.

2.1 Ring of polynomials

Definition. A *polynomial* in one variable with coefficients in any ring R is an expression of the form

$$f(x) = a_n x^n + a_{n-1}x^{n-1} + \cdots + a_o, \text{ where } a_i \in R \; for \; all \; i.$$

Formally, a polynomial $f(x)$ is determined by its vector or coefficients $(a_o, a_1, \cdots, a_n, \cdots)$ where $a_i \in R$ and all but a finite number of a_i's are zero. The *degree* of a non-zero polynomial $f(x)$ (written deg $f(x)$) is the largest integer k such that the coefficient a_k of x^k is not zero. The coefficient of the highest degree of a polynomial which is not zero is called its leading coefficient. A *monic polynomial* is one whose leading coefficient is 1 (the identity element of R).

Addition and multiplication of polynomials mimic the familiar operations on real polynomial functions.

More precisely, if $R[x]$ denotes the set of polynomials with coefficients in R, with the addition $(a_o, a_1, \cdots) + (b_o, b_1, \cdots) = (a_o + b_o, a_1 + b_1, \cdots)$, $(R[x], +)$ is an abelian group.

With multiplication in $R[x]$ defined by

$$(a_o, a_1, \cdots).(b_o, b_1, \cdots) = (p_o, p_1, \cdots), \text{ where} p_i = \sum_{j+k=i} a_j b_k,$$

$(R[x], +, .)$ becomes a commutative ring with $0 = (0, 0, \cdots)$ as the zero element and $1 = (1, 0, \cdots)$ as the identity element.

For defining ring of polynomials in two variables x and y, the reader should be able to observe that the rings $(R[x])[y]$ and $(R[y])[x]$ are naturally isomorphic. Identifying these two rings and writing $R[x, y]$ for them, one calls $R[x, y]$ the *ring of Polynomials in two variables* x and y. Similarly, one defines the ring of polynomials $R[x_1, \cdots, x_n]$ in n variables. For an element $f = f(x_1, x_2, \cdots, x_n) = \sum a_{i_1 \cdots i_n} x_1^{i_1} \cdots x_n^{i_n} \in F[x_1, \cdots, x_n]$ the *degree* of f is the maximum of the sums $i_1 + i_2 + \cdots + i_n$ taken over all the terms.

Exercise 2.1. If R is an integral domain, show that so is the ring $R[x]$.

Exercise 2.2. Let $\phi : R \longrightarrow R'$ be a ring homomorphism. Given an element $\alpha \in R'$, show that there is a unique homomorphism $\Phi : R[x] \longrightarrow R'$ which sends x to α and whose restriction on R is ϕ.

Let R be a subring of R', $\alpha \in R'$ and $i : R \longrightarrow R'$ be the inclusion map. As in the above exercise, we denote the unique homomorphism (called an evaluation map) extending i to $R[x]$ by I.

For $f \in R[x]$, if $f(\alpha) \stackrel{def}{=} I(f) = 0$, then α is said to be a *root* or a *zero* of f in R'.

Let $i : \mathbf{Z} \longrightarrow \mathbf{C}$ be the inclusion map. Now, for any complex number α, if we consider the unique homomorphism $I : \mathbf{Z}[x] \longrightarrow \mathbf{C}$ as in above, then the image, which is denoted by $\mathbf{Z}[\alpha]$, will be the smallest ring of $\mathbf{C}$ containing α. $\mathbf{Z}[\alpha]$ will consist of all elements of the form $a_n\alpha^n + a_{n-1}\alpha^{n-1} + \cdots + a_o, \quad a_i \in \mathbf{Z}$. If the kernel of the map I is not zero, then α is said to be an *algebraic number.* On the other hand, if there is no polynomial with integer coefficients having α as a root, then α is called a *transcendental number.* Since the set of algebraic numbers is countable, it follows that transcendental numbers do exist. If α is a transcendental number, then $\mathbf{Z}[\alpha] \cong \mathbf{Z}[x]$, the ring of polynomials, by $p(\alpha) \mapsto p(x)$.

2.2 Division in polynomial rings

Division with remainder for polynomials reads:

Theorem 2.1. If $f(x), g(x) \in R[x]$ where the leading coefficient of $f(x)$ is a unit in R, then there are polynomials $q(x), r(x) \in R[x]$ such that $g(x) = f(x)q(x) + r(x)$ where $\deg r(x) < \deg f(x)$ or $r(x) = 0$.

Proof: The proof is by 'long division' process. If degree of $g(x)$ is strictly less than that of $f(x)$, then we can put $q(x) = 0$ and $r(x) = g(x)$ to have the equation $g(x) = 0f(x) + g(x)$ satisfying the required conditions.

Therefore, we assume that $g(x)$ and $f(x)$ have degrees m and n respectively with $m \geq n$.

Let $f(x) = a_n x^n + a_{n-1}x^{n-1} + \cdots + a_o$ and $g(x) = b_m x^m + b_{m-1}x^{m-1} + \cdots + b_o$. By assumption, a_n is a unit in R. Hence, $\exists a_n^{-1} \in R$ satisfying $a_n a_n^{-1} = 1$.

We consider the polynomial $g_1(x) = g(x) - b_m a_n^{-1} x^{m-n} f(x)$. Suppose $g_1(x) = c_k x^k + c_{k-1}x^{k-1} + \cdots + c_o, c_k \neq 0$. Then $k \leq m-1$. If $k \geq n$ we repeat the process to get $g_2(x)$ and so on, till we arrive at $g_i(x)$ whose degree is strictly less than n. We observe that $g(x) = (b_m a_n^{-1} x^{m-n} + c_k a_n^{-1} x^{k-n} + \cdots)f(x) + r(x)$ with $r(x) = g_i(x)$. □

Remark 2.1. From the above theorem it follows easily that if $g(x)$ is a polynomial in $R[x]$ and $\alpha \in R$ such that $g(\alpha) = 0$, then $g(x) = (x - \alpha)q(x)$ for some $q(x) \in R[x]$.

The proof that every ideal of the ring of integers is principal (see Exercise 1.1 (c)) can be adopted to prove the next exercise.

Exercise 2.3. Let F be a field. Show that every ideal in the polynomial ring $F[x]$ is a principal ideal.

Exercise 2.4. Let F be a field and $f, g \in F[x]$, not both zero. Then there is a unique monic polynomial $d(x)$ called the greatest common divisor of f and g such that

(a) d generates the ideal (f, g) of $F[x]$ generated by the two polynomials.

(b) d divides f and g.

(c) If h is a divisor of f and g, then h divides d.

(d) There are polynomials $p, q \in F[x]$, such that $d = pf + qg$.

Definition. Let F be a field. A polynomial $p(x) \in F[x]$ is called *irreducible* if it is not constant and its only divisors of lower degree in $F[x]$ are constants.

Usually, one normalizes an irreducible polynomial by factoring out the leading coefficient, so as to make it monic.

As in the case of integers, one obtains the following result.

Theorem 2.2. Let F be a field and $F[x]$ the polynomial ring in one variable over F. Then every non-zero polynomial $f \in F[x]$ can be written as a product $f = cp_1 \cdots p_r$ where $c\ (\neq 0) \in F$ and p_i 's are monic irreducible polynomials in $F[x], r \geq 0$. This factorization is unique, except for the ordering of the terms.

Theorem 2.3. Let F be a field and $f(x)$ a polynomial of degree n with coefficients in F. Then f has at most n roots in F.

Proof: There is nothing to prove if f is of degree 1. Suppose $\alpha \in F$ is a root of f. Then $f(x) = (x - \alpha)q(x)$ where $q(x)$ is a polynomial of degree $n - 1$. If β is another root, then $(\beta - \alpha)q(\beta) = 0$. If $\beta \neq \alpha$, then $q(\beta) = 0$. But by induction $q(x)$ has at most $n - 1$ roots in F. □

Remark 2.2. The fact that F is a field is crucial to the above theorem. For instance, if $R = \mathbf{Z}/8\mathbf{Z}$, the polynomial $x^2 - 1$ in $R[x]$ has four roots modulo 8, as $x^2 - 1 = (x + 1)(x - 1) = (x + 3)(x - 3)$.

Exercise set A

A.1 Prove that for any integer $n \geq 2$, $\frac{1}{2} + \frac{1}{3} + \cdots + \frac{1}{n}$ is not an integer.

A.2 Let a and b be relatively prime integers. Prove that there are integers m, n such that $a^m + b^n \equiv 1 \pmod{ab}$.

A.3 Recall that for a positive integer n, $\phi(n)$ (*Euler's Totient*) is defined to be the number of integers between 1 and n relatively prime to n. Prove that $\sum_{d|n} \phi(d) = n$.

A.4 Let G be a finite group of order n with identity element 1. Suppose that, for any divisor d of n, the number of elements a in G satisfying $a^d = 1$ does not exceed d. Show that G is cyclic.

A.5 Prove that $7 + 2^{3/2}$ and $\sqrt{3} + \sqrt{-5}$ are algebraic numbers.

A.6 Let $\mathbf{Q}\ [\alpha, \beta]$ denote the smallest subring of $\mathbf{C}$ containing $\mathbf{Q}$, $\alpha = \sqrt{2}$ and $\beta = \sqrt{3}$. Let $\gamma = \alpha + \beta$. Prove that

$$\mathbf{Q}\ [\alpha, \beta] = \mathbf{Q}\ [\gamma].$$

A.7 Describe the group of units in the ring $\mathbf{Z}/12\mathbf{Z}$.

A.8 The *ring of Gaussian integers* is a subring of complex numbers defined by $\mathbf{Z}[i] = \{a + ib : a, b \in \mathbf{Z}\}$. Show that the ring $\mathbf{Z}[i]/(1 + i3)$ is isomorphic to the ring $\mathbf{Z}/10Z$.

A.9 Prove that every non-zero ideal in the ring of Gaussian integers contains a non-zero integer.

Chapter 3

Rings and fields revisited

In this chapter, we go through few more definitions and elementary results on rings and fields which could not be done before the chapters on integers and polynomails. We also derive some important number theoretic results with our algebraic tools.

3.1 Characteristic of a ring

Definition. Let R be a ring. If there exists a positive integer n such that $n1 = \underbrace{1 + \cdots + 1}_{n \text{ times}} = 0$, then the least such positive integer is called the *characteristic* of the ring R. In this case we say that the ring R is of *finite characteristic*. If there is no such integer n, then R is said to be (rather paradoxically) of *characteristic zero*.

Exercise 3.1. An integral domain is either of characteristic zero or characteristic p, where p is a prime number.

Remark 3.1. From exercise 1.4, if $(a, m) = 1$, the congruence $ax \equiv 1 \pmod{m}$ has a unique solution modulo m. Hence we observe that if p is a prime, then $\mathbf{Z}/p\mathbf{Z}$ is a field (This field will be denoted by $\mathbf{F}_p$).

$\mathbf{F}_p$ has characteristic p. The ring $\mathbf{Z}$ and the field $\mathbf{Q}$ are of characteristic zero.

As the characteristic of a ring R is nothing but the nonnegative integer which generates the kernel of the homomorphism from $\mathbf{Z}$ which takes 1 to the identity element of R, it follows that R will contain either a subring isomorphic to $\mathbf{Z}$ or $\mathbf{Z}/m\mathbf{Z}$ for some positive integer m. If R is an integral domain, then it will contain a subring isomorphic to $\mathbf{Z}$ or $\mathbf{Z}/p$ for a prime p. Therefore, every field will contain either a subfield isomorphic to $\mathbf{Q}$ or a subfield isomorphic to $\mathbf{F}_p$ for a prime p.

Definition. The fields $\mathbf{Q}$ and F_p are called *prime fields.* A field having finitely many elements is called a *finite field.*

Remark 3.2. The fields $\mathbf{F}_p$ already provide us with an infinite supply of finite fields. We shall see that there are finite fields other than these. A finite field will definitely have finite characteristic. The polynomial ring $\mathbf{F}_p[x]$ is an example of an infinite integral domain having finite characteristic.

Remark 3.3. From Exercise A.4 and Theorem 2.3, it follows that a finite subgroup of the multiplicative group of non-zero elements of a field has to be cyclic. In particular, the multiplicative group of non-zero elements of a finite field is cyclic.

Exercise 3.2. If G is a finite abelian group, then $\exists x \in G$ whose order is the least common multiple of the orders of the elements of G. Hence give another proof of the fact that the multiplicative group of non-zero elements of a finite field is cyclic.

Exercise 3.3. Let R be an integral domain of characteristic p where p is a prime. Prove the map $R \longrightarrow R$ defined by $x \mapsto x^p$ is a ring homomorphism. (This is called the *Frobenius homomorphism).*

3.2 Wilson's theorem

Theorem 3.1.

i) (**Wilson's theorem**) For any prime number p,
$$(p-1)! \equiv -1 \pmod{p}.$$

ii) The converse of Wilson's theorem is also true. That is, if a positive integer $n > 1$ divides $(n-1)! + 1$, then n is a prime number.

iii) Let p be a prime number of the form $4n+1$. Using Wilson's theorem, prove that there is a natural number n such that $n^2 + 1$ is divisible by p.

Proof: We consider the mutiplicative group of non-zero elements of the field $\mathbf{F}_p$. For an integer a, $\overline{a}$ will denote its residue class modulo p. Since the only roots of the equation $x^2 - 1 = 0$ in $\mathbf{F}_p$ are $\pm\overline{1}$, there is no other non-zero element which is its own (multiplicative) inverse. Therefore, all other non-zero elements in F_p can be paired as (α, α^{-1}).

Thus $\overline{1}\,.\overline{2}\,.\cdots\,\overline{p-1} = \overline{1}\,.\overline{(-1)} = \overline{(-1)}$

This completes the proof of part (i).

To prove (ii), let $n > 1$ be a positive integer which divides $(n-1)! + 1$. Then $nk = (n-1)! + 1$ for some integer k and this implies that none of the integers $1, \cdots, (n-1)$ divides n. Hence n is a prime number.

Now, let p be a prime number of the form $4n+1$. From (i), we have
$\overline{(-1)} = \overline{1}.\,\overline{2}\cdots\,\overline{(p-1)}$
$= \left(\overline{1}\,.\overline{2}\cdots\,\overline{((p-1)/2)}\right)\left(\overline{(-(p-1)/2)}\cdots\overline{(-2)(-1)}\right)$
Therefore, p divides $(-1)^{(p-1)/2}(1.2.\cdots\frac{p-1}{2})^2 + 1$.
Since p is of the form $4n+1$, $(-1)^{(p-1)/2} = 1$ and we are through. □

Exercise 3.4. From Theorem 2.3, deduce that for a prime p, each of the coefficients of the polynimial $f(x) = (x-1)(x-2)\cdots(x-p+1) - x^{p-1} + 1$ is divisible by p. Observe that this yields another proof of Wilson's theorem.

3.3 A result on vector spaces

We end this chapter with an interesting result for vector spaces over infinite fields. We shall be using this result later on.

Theorem 3.2. Let V be a vector space over an infinite field K. Then V can not be written as a union of a finite number of proper subspaces.

Proof: The proof is by induction on the number n of subspaces. If $n = 1$, then the result is obvious. We assume the result for $n < m$. Now, suppose we have m proper subspaces $V_1, \cdots, V_m$.

By the induction hypothesis, there is $e \in V$ such that $e \notin V_i, i = 1, \cdots, m-1$. If $e \notin V_m$, there is nothing to prove. Suppose $e \in V_m$. We choose $f \notin V_m$. Then $e + cf \notin V_m$ for all non-zero $c \in K$.

We claim that there is $c_0 \in K^*$ such that $e + c_0 f \notin V_i$ for all $i, 1 \leq i \leq m$. Otherwise, since K is infinite, there exist $c_1, c_2 \in K^*, c_1 \neq c_2$ such that

$e + c_1 f, e + c_2 f \in V_i$ for some $i < m$. Then $(c_1 - c_2)f \in V_i$, that is $f \in V_i$ and hence $e \in V_i$, which is a contradiction. □

Chapter 4

Factorization

In this chapter, we consider the problem of factorization in integral domains. In chapter 1, we have seen that in the ring of integers, the fundamental theorem of arithmetic assures unique factorization of non zero elements into product of primes upto arrangements and multiples of ± 1. One looks for rings with similar properties. In the case of integers, the division algorithm was used to prove unique factorization. One generalizes this division algorithm in a suitable way and looks for rings satisfying such properties. We shall see that for a field F, the polynomial ring $F(x)$ will be one such candidate. We shall see some other examples as well.

4.1 Divisibility

We start with generalizing the concept of divisibility which was introduced in the case of integers in Chapter 1.

Definition. Let R be an integral domain. A non-zero element $a \in R$ is said to *divide* another element b if $b = aq$ for some $q \in R$ and we write $a|b$.

The element a is a *proper divisor* of b if $b = aq$ for some $q \in R$ and neither a nor q is a unit.

A non-zero element a of R is called *irreducible* if it is not a unit and it has no proper divisor.

Two elements $a, a' \in R$ are called *associates* if each divides the other, in other words if there is some unit u such that $a = ua'$.

The following statements are obvious.

1) u is a unit if and only if $(u) = (1)$.

2) a, a' are associates if and only if $(a) = (a')$.

3) a divides b if and only if $(b) \subset (a)$.

4) a is a proper divisor of b if and only if $(b) < (a) < (1)$.

4.2 UFD and PID

Theorem 4.1. Let R be an integral domain. Then the following conditions are equivalent.

(a) For any $a \in R$ which is non-zero and non-unit, the process of factoring a terminates after finitely many steps and results in a factorization $a = b_1 b_2 \cdots b_r$ of a into irreducible elements of R.

(b) R does not contain an infinte increasing chain of principal ideals $(a_1) < (a_2) < \cdots < (a_n) < \cdots$.

Proof: $(a) \Longrightarrow (b)$:

Suppose (a) holds. If possible, let R contain an infinite ascending chain of principal ideals $(a_1) < (a_2) < \cdots < (a_n) < \cdots$. Clearly none of the a_i's are units. Now $(a_n) < (a_{n+1})$. This implies a_{n+1} is a proper divisor of a_n, say $a_n = a_{n+1} b_{n+1}$ where a_{n+1}, b_{n+1} are not units. This gives a non terminating process of factorization of a_1, viz,

$$a_1 = a_2 b_2 = a_3 b_3 b_2 = \cdots\cdots = a_n b_n b_{n-1} \cdots b_2 = \cdots\cdots\cdots$$

which is a contradiction.

$(b) \Longrightarrow (a)$:

Clearly, a sequence of non-terminating process of factorization gives us an increasing chain of principal ideals. □

Definition. An integral domain R is called a *factorization domain* if any element $r \in R$ has a factorization into irreducible elements.

Example. Let x_k be the 2^k-th root of x_1 in some extension of the quotient field $F(x_1)$ of the polynomial ring $F[x_1]$, where F is a field.

Let $R = F[x_1, x_2, \cdots]$. Now, $x_1 = x_2^2 = (x_3^2)^2 = \cdots\cdots$

gives a non-terminating process of factorization of x_1. In other words, R is not a factorization domain.

Remark 4.1. It should be remarked that the situation as in the above is not encountered often. Usually, factorization of a non-zero element into irreducible elements becomes possible, but it is not unique.

For instance, consider the integral domain $R = \mathbf{Z}[\sqrt{-5}]$. It consists of complex numbers of the form $a + b\sqrt{-5}$ where $a, b \in \mathbf{Z}$. We shall study this ring in detail in later chapters. One observes that the units in this ring are ± 1 and the element $6 \in R$ has two essentially different factorizations in R, viz,

$$6 = 2 \times 3 = (1 + \sqrt{-5})(1 - \sqrt{-5}).$$

Definition. Let R be an integral domain. An element $p \in R$ will be called *prime* if p is neither zero nor a unit and if p divides a product of elements of R, then it divides one of the factors.

An integral domain R is called a *unique factorization domain* (UFD) if it has the following two properties:

(i) The process of factoring a non-zero element a which is not a unit terminates after finitely many steps and yields a factorization $a = p_1 p_2 \cdots p_m$, where p_i's are irreducible elements of R.

(ii) If a is factored in two ways into irreducible elements, say

$$a = p_1 p_2 \cdots p_m = q_1 q_2 \cdots q_n$$

then, $m = n$ and $q_1, q_2, \cdots, q_n$ can be rearranged as $q_{i_1}, q_{i_2}, \cdots, q_{i_m}$ such that p_j is associate of q_{i_j}, for all j.

The proof of the following theorem uses arguments already seen (Theorems 1.4, 1.5).

Theorem 4.2. Let R be an integral domain. Suppose that existence of factorization holds in R. Then R is a UFD if and only if every irreducible element is prime.

Exercise 4.1. Give an example of a ring R where R contains an element a which is prime, but not irreducible.

Exercise 4.2. Prove that in an integral domain, a prime element is irreducible.

Exercise 4.3. Give an example of an integral domain D, where D contains an element a which is irreducible, but not prime.

Exercise 4.4. Prove that in a principal ideal domain, an irreducible element is prime.

Exercise 4.5. Let R be a unique factorization domain and let a, b be elements of R which are not both zero. Then there exists a greatest common divisor d of a, b with the following properties:

(i) d divides a and b.

(ii) if an element e of R divides a and b, then e divides d.

Theorem 4.3. A PID is a UFD.

Proof: Let R be a PID. Then, by Exercise 4.5, every irreducible element of R is prime. Therefore by Theorem 4.2, it is enough to prove the existence of factorizations for R, which further is equivalent to showing that R contains no infinite increasing chain of principal ideals.

If possible, suppose $(a_1) < (a_2) < \cdots < (a_n) < \cdots$ be an infinite increasing chain of principal ideals in R. Now, by the above observation, the union I of the above chain is an ideal of R. Again, since R is a PID, $I = (b)$ for some $b \in R$. Since $b \in I$, $b \in (a_n)$ for some n. This implies $(b) \subset (a_n)$.

On the other hand, $(a_n) \subset (a_{n+1}) \subset (b)$. Therefore we have $(a_n) = (a_{n+1}) = (b)$, which contradicts the fact that $(a_n) < (a_{n+1})$. This completes the proof of the theorem. □

Exercise 4.6. Let R be a PID which is not a field. Then any proper ideal A of R is a maximal ideal iff it is generated by an irreducible element of R.

Theorem 4.4. Let R be a PID and p a non-zero element of R. Then $R/(p)$ is a field iff p is irreducible.

Proof: Let p be irreducible. Then, the only principal ideals which contain the ideal (p) are (p) and (1). Then the ideal (p) is maximal which further implies that $R/(p)$ is a field.

Conversely, let $b \in R$ have a proper factorization $b = aq$, where neither a nor q is a unit. Then $(b) < (a) < (1)$. This implies the ideal (b) is not maximal and hence $R/(b)$ is not a field. □

4.3 Euclidean domains

Definition. A *size function* on an integral domain R is a function

$$\sigma : R \setminus \{0\} \longrightarrow \mathbf{N},$$

where $\mathbf{N}$ is the set of non-negative integers.

Examples. We have the size functions in the rings $\mathbf{Z}$ and $F[x]$ given by the absolute value and the degree respectively. In the ring $\mathbf{Z}[i]$ of Gaussian integers (introduced in Erercise A.8) also we have a size function which is given by (absolute value)^2.

An integral domain is said to be a *Euclidean domain* if there is a size function σ on R which satisfies the division algorithm:

If $a, b \in R$ and $a \neq 0$, there exist $q, r \in R$ such that $b = aq + r$ when either $r = 0$ or $\sigma(r) < \sigma(a)$.

Theorem 4.5. The ring $\mathbf{Z}$, the polynomial ring $F[x]$ over a field F and the ring $\mathbf{Z}[i]$ are Euclidean domains.

Proof: In Theorem 1.1, we have already proved the result for $\mathbf{Z}$. Again, since every non-zero element of a field is invertible, by Theorem 2.1, the result is established for the case $F[x]$ as well.

So, we consider the ring $\mathbf{Z}[i]$ with the size function σ defined by $\sigma(x) = |x|^2$. Let $a, b \in \mathbf{Z}[i]$ with $a \neq 0$. Let $b = a\omega$ where $\omega = x + iy$ is a complex number. Now there is a Gaussian integer $m + in$ such that $x = m + x_o, y = n + y_o$ where $-\frac{1}{2} \leq x_o, y_o \leq \frac{1}{2}$.

Now,

$$|b - (m + in)a|^2 = |(x_o + iy_o)a|^2 < \frac{1}{2}|a|^2$$

which completes the proof.

The proof of the following theorem is now routine.

Theorem 4.6. A Euclidean domain is a PID (and hence a UFD).

Corollary 4.6.1. The rings $\mathbf{Z}$, $F[x]$ (when F is a field) and $\mathbf{Z}[i]$ are principal ideal domains.

Remark 4.2. In proving Theorem 1.2 on the existence of greatest common divisors, the property of the ring of integers which played the crucial role is that the absolute value is a size function which makes it a Euclidean domain. Thus the next exercise is just to make things work in the general situation.

Exercise 4.7. In an Euclidean domain R, any two elements a and b, where one of them, say a, is non-zero have a greatest common divisor d. Moreover, $d = \lambda a + \mu b$ for some $\lambda, \mu \in R$.

Chapter 5

Gauss lemma and Eisenstein criterion

In this chapter, we mainly dwell on some questions related to the polynomial rings $\mathbf{Z}[x]$ and $\mathbf{Q}[x]$. From Theorem 4.5 of the previous chapter, we know that for a field F, the polynomial ring $F[x]$ is a Euclidean domain and hence a UFD (see Theorem 4.6). In this chapter, we shall see that for $R[x]$ to be a UFD, it is enough that R is a UFD.

5.1 Gauss lemma

We start with the following definition.

Definition. Let $f(x) = a_n x^n + \cdots + a_o \in \mathbf{Z}[x]$. Then $f(x)$ is called *primitive* if its coefficients $a_o, \cdots, a_n$ have no common integer factor except for the units ± 1 of $\mathbf{Z}$ and if the leading coefficient a_n is positive.

Exercise 5.1. Every non-zero polynomial $f(x) \in \mathbf{Q}[x]$ can be written as a product $f(x) = cf_o(x)$ where $c \in \mathbf{Q}$ and $f_o(x) \in \mathbf{Z}[x]$ is primitive. Moreover, this expression for $f(x)$ is unique.

Remark 5.1. It is clear that $f(x)$ has integer coefficients if and only if c is an integer. In that case, then $|c| > 0$ is a greatest common divisor of the coefficients of $f(x)$ and the sign of c is the sign of the leading coefficient of $f(x)$.

Definition. The rational number c which appears in Exercise 5.1 is called the *content* of $f(x)$. If $f(x)$ has integer coefficients, then the content divides $f(x)$ in $\mathbf{Z}[x]$; $f(x)$ is primitive iff its content is 1.

Theorem 5.1. (Gauss Lemma) Product of two primitive polynomials in $\mathbf{Z}[x]$ is primitive.

Proof: Let $f(x), g(x)$ be two primitive polynomials in $\mathbf{Z}[x]$. Let $h(x) = f(x)g(x)$. The only thing we have to show is that no prime integer p divides all the coefficients of $h(x)$.

We consider the homomorphism $\mathbf{Z}[x] \longrightarrow \mathbf{F}_p[x]$ given by $f(x) = a_m x^m + \cdots + a_o \mapsto \overline{f}(x) = \overline{a}_m x^m + \cdots + \overline{a}_o$ where the coefficients are read modulo p. Since, $f(x)$ and $g(x)$ are primitive, $\overline{f} \neq 0$ and $\overline{g}(x) \neq 0$. Now, we invoke the fact that the polynomial ring $\mathbf{F}_p[x]$ is an integral domain. This implies that $\overline{h}(x) = \overline{f}(x)\overline{g}(x) \neq 0$. Hence $h(x)$ is primitive. □

Corollary 5.1.1. A finite product of primitive polynomials in $\mathbf{Z}[x]$ is again primitive.

Theorem 5.2. Suppose $f(x), g(x) \in \mathbf{Z}[x]$ with f being primitive. If $g(x) = f(x)q(x)$ where $q(x) \in \mathbf{Q}[x]$, then $q(x) \in \mathbf{Z}[x]$.

Proof: Let $q = cq_o$ where $q_o \in \mathbf{Z}[x]$ is primitive and $c \in \mathbf{Q}$. By Gauss lemma, $fq_o$is primitive. Now $g = cfq_o$ shows that $fq_o = g_o$, the primitive polynomial associated to g. Since $g \in \mathbf{Z}[x]$, it follows that $c \in \mathbf{Z}$; in other words, $q \in \mathbf{Z}[x]$. □

Corollary 5.2.1. Let $f(x), g(x) \in \mathbf{Q}[x]$ and $f_o(x), g_o(x)$ be the associated primitive polynomials in $\mathbf{Z}[x]$. If $f(x)$ divides $g(x)$ in $\mathbf{Q}$ $[x]$, then $f_o(x)$ divides $g_o(x)$ in $\mathbf{Z}[x]$.

Proof: If $f(x)$ divides $g(x)$ in $\mathbf{Q}$ $[x]$, then clearly $f_o(x)$ divides $g_o(x)$ in $\mathbf{Q}[x]$.

Let $g_o(x) = q(x)f_o(x)$, where $q(x) \in \mathbf{Q}[x]$. By Theorem 5.2, $q(x) \in \mathbf{Z}[x]$ and we are through. □

Corollary 5.2.2. Let $f(x), g(x) \in \mathbf{Z}[x]$. Suppose that $f(x), g(x)$ have a common non-constant factor $h(x)$ in $\mathbf{Q}[x]$. Then they have a common non-constant factor in $\mathbf{Z}[x]$ too.

Proof: If $h_o(x)$ is the primitive polynomial associated to $h(x)$, then $h_o(x)$, also divides $f(x)$ and $g(x)$ in $\mathbf{Q}[x]$. By Theorem 5.2, $h_o(x)$ divides $f(x)$ and $g(x)$ in $\mathbf{Z}[x]$. □

Corollary 5.2.3. If a non-constant polynomial $f(x)$ is irreducible in $\mathbf{Z}[x]$, then it is irreducible in $\mathbf{Q}[x]$.

Theorem 5.3. Let $f(x) \in \mathbf{Z}[x]$ be such that its leading coefficient is positive. Then $f(x)$ is irreducible in $\mathbf{Z}[x]$ if and only if either

(1) $f(x)$ is a prime integer, or

(2) f is a primitive polynomial which is irreducible in $\mathbf{Q}[x]$.

Proof. Let $f(x)$ be irreducible. Let $f(x) = cf_o(x)$ where $f_o(x)$ is primitive. Since $f(x)$ is irreducible, either c or $f_o(x)$ is 1. If $f_o(x) = 1$, then $f(x)$ is a constant and hence a prime integer. If $c = 1$, then $f(x)$ is primitive. Also by Corollary 5.2.3, it is irreducible in $\mathbf{Q}[x]$. The converse is obvious. □

Theorem 5.4. In $\mathbf{Z}[x]$, every irreducible element is a prime element.

Proof: Let $f(x) \in \mathbf{Z}[x]$ be irreducible. Let $f(x) \mid g(x)h(x)$, where $g(x), h(x) \in \mathbf{Z}[x]$.

Case I. ($f(x) = p$ is a prime integer.)

Let $g(x) = cg_o(x), h(x) = dh_o(x)$ where $g_o(x)$ and $h_o(x)$ are the primitive polynomials associated to $g(x)$ and $h(x)$ respectively. By the Gauss lemma, $g_o(x)h_o(x)$ is primitive and therefore some coefficients, say, a of $g_o(x)h_o(x)$ is not divisible by p. But since p divides $g(x)h(x)$, $p \mid cda$ which implies $p \mid c$ or d, which would imply that $p \mid g(x)$ or $p \mid h(x)$.

Case II. ($f(x)$ is a primitive polynomial which is irreducible in $\mathbf{Q}[x]$.)

As we have already seen, $\mathbf{Q}[x]$ is an Euclidean domain and hence a PID. Therefore $f(x)$ is a prime element of $\mathbf{Q}[x]$. Hence $f(x)$ divides $g(x)$ or $h(x)$ in $\mathbf{Q}[x]$. Therefore, by Theorem 5.2, $f(x)$ divides $g(x)$ or $h(x)$ in $\mathbf{Z}[x]$. □

Theorem 5.5. The polynomial ring $\mathbf{Z}[x]$ is a UFD.

Proof: Given a non-zero non-unit $f(x)$ in $\mathbf{Z}[x]$, considering the factorization of $f(x)$ in $\mathbf{Q}[x]$ and clearing denominators, existence of factorization can be proved for $\mathbf{Z}[x]$. Therefore, by Theorem 4.2 and Theorem 5.4, the result follows. □

Proceeding along the same line, one can prove the following more general statement.

Theorem 5.6. If D is a UFD, then $D[x]$ is a UFD.

Remark 5.2. It follows from the above theorem that the polynomial rings $\mathbf{Z}[x_1, x_2, \cdots, x_n]$ and $F[x_1, x_2, \cdots, x_n]$ where F is a field, are UFD's.

5.2 Eisenstein criterion

Theorem 5.7. (Eisenstein criterion) Let $f(x) = a_n x^n + \cdots + a_o \in \mathbf{Z}[x]$ be an integer polynomial. Let p be a prime integer such that p divides $a_{n-1}, a_{n-2}, \cdots, a_o$ and $p \nmid a_n, p^2 \nmid a_o$. Then $f(x)$ is irreducible in $\mathbf{Q}[x]$.

Proof: Suppose $f(x)$ is reducible in $\mathbf{Z}[x]$ into factors of positive degree, say $f(x) = g(x)h(x)$. Reducing modulo p, $\overline{f}(x) = \overline{a}_n x^n$ where $\overline{a}_n \neq 0$.

Now, $\overline{g}(x) \mid \overline{f}(x), \overline{h}(x) \mid \overline{f}(x)$ and hence $\overline{g}(x)$ and $\overline{h}(x)$ are monomials. Therefore all coefficients of $g(x)$ and $h(x)$ except the leading coefficients, are divisible by p. Let the constant coefficients of $g(x), h(x)$ be b_o, c_o respectively. This implies that $a_o = b_o c_o$. But $p \mid b_o$ and $p \mid c_o$ and this implies $p^2 \mid a_o$, which is a contradiction. Hence, $f(x)$ is irreducible in $\mathbf{Z}[x]$ and hence so in $\mathbf{Q}[x]$. □

Exercise 5.2. Show that $f(x) = 8x^3 - 6x - 1$ is irreducible in $\mathbf{Q}[x]$.

Chapter 6

Field extensions

In this chapter, we dwell briefly on some special classes of finite extension fields and prove a very useful result (Theorem 6.3) on finite separable extensions. Towards the end of the chapter, we give some results of arithmetical nature related to the number of solutions of polynomials over finite fields.

6.1 Algebraic extensions

Definition. Let K be an extension field of F. Let $\alpha_1, \alpha_2, \cdots, \alpha_r \in K$. Then the smallest subfield of K containing F and the α_i's is denoted by $F(\alpha_1, \cdots, \alpha_r)$ and is said to be *generated by* the α_i's over F. An extension K of F is called a *simple extension* if it is generated by a single element over F, that is, if $K = F(\alpha)$ for some $\alpha \in K$.

Let K be an extension field of a field F and $\alpha \in K$. We consider the evaluation map $I : F[x] \to F(\alpha)$ given by $g(x) \mapsto g(\alpha)$. As in the case of defining algebraic and transcendental numbers, if the kernel of I, $\ker(I)$ is zero, then α is said to be *transcendental* over F. Otherwise, α is said to be *algebraic* over F.

If α is algebraic over F, the non-zero ideal $\ker(I)$ of $F[x]$ will be a principal ideal generated by a polynomial, say $f(x)$. It is easy to see that $f(x)$ is irreducible. Dividing by the leading coefficient, if necessary, we can assume $f(x)$ to be monic. This monic irreducible polynomial $f(x)$ is called the *minimal polynomial* of α over F.

If K is an extension field of F, then K is a vector space over F. The dimension of K as a vector space over F is called the *degree* of K over F and is denoted by $[K : F]$. If $[K : F]$ is finite, K is called a *finite extension* of F. If every element of K is algebraic, then K is called an *algebraic extension* of F.

Exercise 6.1. Show that a finite extension K of a field F must be an algebraic extension of F.

Exercise 6.2. Let K be an extension field of a field F and $\alpha \in K$ be algebraic over F. If n is the degree of the minimal polynomial of α over F, then show that $[k(\alpha) : k] = n$. (In this case, α is said to be *algebraic of degree* n over F.)

Exercise 6.3. Let K be a finite extension of a field F and L be a finite extension K. Show that $[L : F] = [L : K][K : F]$.

Remark 6.1. Let K be an algebraic extension of F and L an algebraic extension of K. We observe that by Exercises 6.1–6.3, it follows that L is an algebraic extension of F. For, let α be any element of L. Then there is a non-zero polynomial $f(x) = a_n x^n + \cdots + a_0$ in $K[x]$ such that $f(\alpha) = 0$. Let $K_1 = F(a_n, \cdots, a_0)$ be the field generated by $\{a_n, \cdots, a_0\}$ over F. Now, α being algebraic over K_1, $K_1(\alpha)$ is finite over K_1 by Exercise 6.2. Again, by Exercises 6.2 and 6.3, K_1 is a finite extension of F. Therefore, $K_1(\alpha)$ is finite over F and by Exercise 6.1, α is algebraic over F.

Definition. Let K be an extension field of a field F. The set of all elements of K which are algebraic over F is called the *algebraic closure* of F in K.

If α and β are two elements of K which are algebraic over F, then $F(\alpha, \beta)$, the smallest subfield of K containing F, α and β, is a finite extension of F and hence an algebraic extension by Exercise 6.1. Since $\alpha \pm \beta, \alpha\beta \in F(\alpha, \beta)$, these elements are algebraic over F. If $\alpha \neq 0$, the same is true about α^{-1}. Hence the algebraic closure of F in K is a subfield of K. If it happens to be equal to F, then F is said to be *algebraically closed* in K.

Remark 6.2. The algebraic closure of $\mathbf{Q}$ in $\mathbf{C}$ will be denoted by $\overline{\mathbf{Q}}$. We observe that $\overline{\mathbf{Q}}$ is a field extension of $\mathbf{Q}$ which is an algebraic extension, but not a finite extension. For, given any positive integer n, the polynomial $x^n - 2$ is irreducible in $\mathbf{Q}[x]$ by Eisenstein criterion (see Theorem 5.7). Hence, if $\alpha \in \overline{\mathbf{Q}}$ is a root of it, $\mathbf{Q}(\alpha)$ ($\subset \overline{\mathbf{Q}}$) is an extension of degree n of F.

6.2 Normal extensions

Theorem 6.1. Let F be a field and $f(x) \in F[x]$ an irreducible polynomial of degree n. Then, there is an extension K of F with $[K : F] = n$ containing a root of $f(x)$.

Proof: By Theorem 4.4, $F[x]/(f(x))$ is a field. The map $a \mapsto (f(x)) + a$ is an injective homomorphism from F to $F[x]/(f(x))$. Therefore, $F[x]/(f(x))$ contains an isomorphic copy of F and can be considered as an extension of F. Consider the natural map $g(x) \mapsto (f(x)) + g(x)$ from $F[x] \to F[x]/(f(x))$. The image α of x in $F[x]/(f(x))$ under this map clearly satisfies $f(\alpha) = 0$. Hence, $K = F[x]/(f(x))$ is the desired extension. We leave it as an exercise to check that $[K : F] = n$. □

Definition. A finite extension K of a field F is said to be a *splitting field* for a polynomial $f(x) \in F[x]$, if $f(x)$ can be factored into a product of linear factors in $K[x]$ but not in $K_1[x]$ for any proper subfield K_1 of K.

Exercise 6.4. Let F be a field and $f(x) \in F[x]$ be of degree $n \geq 1$. Show that splitting field for $f(x)$ exists and is of degree at most $n!$.

Definition. Let K_1 and K_2 be two field extensions of a field F. An isomorphism from K_1 onto K_2, which fixes the elements of F pointwise, is called an *F-isomorphism* and the fields K_1 and K_2 are called *F-isomorphic*. If $K_1 = K_2 = K$, then the F-automorphisms of K form a group under composition 'o' of maps. It is called the *Galois group* of K over F and is denoted by $\mathrm{Gal}(K/F)$.

Let K be an algebraic extension of F. Two elements α_1 and α_2 of K are said to be *conjugates* over F if there is an F-isomorphism σ from $F(\alpha_1)$ onto $F(\alpha_2)$ such that $\sigma(\alpha_1) = \alpha_2$.

Exercise 6.5. Suppose K is an algebraic extension of F and α_1 and α_2 are two elements of K. Show that α_1 and α_2 are conjugates over F if and only if they have the same minimal polynomial over F.

Let F_1 and F_2 be two fields and σ an isomorphism from F_1 onto F_2. For $f(x) = a_n x^n + \cdots + a_0 \in F(x)$, the map $f(x) \mapsto \sigma(a_n)x^n + \cdots + \sigma(a_0)$ is the unique homomorphism (see Exercise 2.2) $F_1[x] \to F_2[x]$ which extends σ. By abuse of notation, we shall write σ for this extension as well.

Exercise 6.6. With notations as above, if K_1 and K_2 are splitting fields of $f(x)$ and $\sigma(f(x))$ respectively over F_1 and F_2, then show that there is an isomorphism from K_1 onto K_2 whose restriction to F_1 is σ.

Remark 6.3. From the above exercise, it follows that two splitting fields of a polynomial over a field F are F-isomorphic. Hence, while talking of splitting field of a polynomial we can say 'the splitting field'.

Exercise 6.7. Let F be a field and K the splitting field of $f(x) \in F[x]$. Let L be any extension field of K. Show that any F-isomorphism $\sigma : K \to L$ maps K onto K itself.

Definition. Let F be a field. A *normal extension* of F is an algebraic extension K of F such that given any irreducible polynomial $f(x) \in F[x]$ having a root in K, $f(x)$ is a product of linear factors in $K[x]$.

Exercise 6.8. Show that a finite normal extension of a field F is nothing but the splitting field of a polynomial $f(x)$ over F.

6.3 Separable extensions

Definition. Let F be a field and $f(x) \in F[x]$. Let K be the splitting field of $f(x)$ over F. If α is a root of $f(x)$ in K, then the *multiplicity* of α is the largest integer n such that $(x - \alpha)^n$ divides $f(x)$ in $K[x]$. A root α of $f(x)$ is said to be a *multiple root* if $n > 1$.

If $f(x) \in F[x]$ is an irreducible polynomial in $F[x]$, then it is called *separable* if it does not have any multiple root.

Let K be an algebraic extension of F. An element $\alpha \in K$ is said to be *separable* over F if the minimal polynomial of $f(x)$ over F is separable. If all the elements of K happens to be separable over F, then K is called a *separable extension* of F. If an extension K of F is not separable, then it is called *inseparable*.

Exercise 6.9. Let $f(x) = a_n x^n + \cdots + a_1 x + a_0 \in F[x]$. One defines the formal derivative $f'(x)$ of $f(x)$ by $f'(x) = na_nx^{n-1} + \cdots + a_1$. Check that this formal derivative satisfies the following properties of derivative as seen in the elementary calculus:

i) If $f(x), g(x) \in F[x]$, and $F(x) = f(x) + g(x)$, then $F'(x) = f'(x) + g'(x)$.

ii) If $f(x), g(x) \in F[x]$, and $F(x) = f(x)g(x)$, then $F'(x) = f'(x)g(x) + g'(x)f(x)$.

Exercise 6.10. Let K be an algebraic extension of F. Show that an element $\alpha \in K$ is not separable over F if and only if $f'(x)$ is the zero polynomial, where $f(x)$ is the minimal polynomial of α over F. Hence deduce that if F is a field of characteristic zero, then any algebraic extension K of F is separable. If F is field of characteristic p for a prime p, show that a polynomial $f(x) \in F[x]$ can have multiple root only if $f(x) = g(x^p)$ for some $g(x) \in F[x]$.

Theorem 6.2. Let K be a finite separable field extension of a field F and $[K : F] = n$. Let N be an extension field of K such that N is a normal extension of F. Then, there are precisely n distinct F-isomorphisms of K into N.

Proof: We proceed by induction on n. If $n = 1$, there is nothing to prove. Let $n > 1$. Let $\alpha \in K$ be such that $\alpha \notin F$. Now, $[K : F(\alpha)] < n$ and K is separable over $F(\alpha)$. Therefore by the induction hypothesis, there are precisely $s = [K : F(\alpha)]$ distinct $F(\alpha)$-isomorphisms σ_i, $i = 1, \cdots, s$, of K into N. Again, since α is separable over F, its minimal polynomial over F will have distinct roots and therefore by Exercise 6.5 there are precisely $t = [F(\alpha) : F]$ distinct F-isomorphisms τ_j, $j = 1, \cdots, t$, of $F(\alpha)$ into N.

N being a splitting field, the F-isomorphisms τ_j's can be extended (see Exercise 6.6) to automorphisms of N whose restrictions on K will be isomorphisms of K into N. We use the same notations τ_j's to denote them. Now, the compositions $\tau_j \ o \ \sigma_i$'s are isomorphisms of K into N. If $\tau_j \ o \ \sigma_i(a) = \tau_u \ o \ \sigma_v(a)$ for all $a \in K$, then for all $a \in F(\alpha)$, we have $\tau_j(a) = \tau_u(a)$ which further implies that $j = u$. Therefore, $\sigma_i(a) = \sigma_v(a)$ for all $a \in K$ and hence $i = v$. Thus, we have proved that $\tau_j \ o \ \sigma_i$ are $s \cdot t$

distinct F-isomorphisms of K into N. That any F-isomorphism of K into N is one among these $s \cdot t$ isomorphisms is not very difficult to see. We leave that as an exercise.

Hence, the number of distinct F-isomorphisms of K into N is $s \cdot t = [K : F(\alpha)] \cdot [F(\alpha) : F] = [K : F] = n$. □

Remark 6.4. The converse of the above theorem is also true. Namely, if K is a finite field extension of a field F with $[K : F] = n$ such that for any normal extension N of F, there are precisely n distinct F-isomorphisms of K into N, then K is a separable extension of F.

Theorem 6.3. Let K be a finite separable extension of F. Then, K is a simple extension. That is, there exists an $\alpha \in K$ such that $K = F(\alpha)$.

Proof:

Case-I (F is a finite field)

Being a finite extension of a finite field, K itself is a finite field and hence by Remark 3.3, K^* is a cyclic group generated by α, say. Clearly, $K = F(\alpha)$.

Case-II (F is an infinite field)

Let $[K : F] = n$. Let N/F be a finite normal extension containing K as a subfield. Such an extension always exists. For, K being a finite extension of F will be generated by some set of finitely many elements $\alpha_1, \alpha_2, \cdots, \alpha_r$ over F. If $f_i(x)$ is the minimal polynomial of α_i over F, then the splitting field of $\prod_{i=1}^{n} f_i(x)$ over F will be one such field. Since K/F is finite separable, by Theorem 6.2 above, there exist n distinct F-isomorphisms $\sigma_1, \cdots \sigma_n$ of K into N. For each $i \neq j$, let $V_{ij} = \{x \in K : \sigma_i(x) = \sigma_j(x)\}$. Then V_{ij} is clearly a subspace of the F-vector space K and since the σ_i's are distinct, it is a proper subspace of K. Therefore by Theorem 3.2, $\cup_{i \neq j} V_{ij}$ is a proper subset of K. This means that there is $\alpha \in K$ such that $\sigma_i(\alpha) \neq \sigma_j(\alpha)$ for $i \neq j$. Therefore α has n distinct conjugates and hence $K = F(\alpha)$. □

6.4 Finite fields

Before we end this chapter, we will have a small discussion on finite fields.

In Remark 3.2 we had observed that a finite field K will contain the field $\mathbf{F}_p$ for some prime p. If $[K : \mathbf{F}_p] = r$, then K will have p^r elements by vector space consideration. We write $q = p^r$. The non-zero elements of K form a cyclic group of order $q-1$ under multiplication. Since the non-zero elements satisfy the equation $x^{q-1} = 1$, we see that the equation $x^q = x$ is satisfied by all the elements of K. This shows that (see Remark 2.1) the polynomial $x^q - x$ factors as $x^q - x = \prod_{a \in K}(x-a)$ in $K[x]$. Clearly, K is the splitting field of $x^q - x$.

From the above discussion, any two finite fields with same number of elements are splitting fields of the same polynomial and hence are isomorphic.

We now claim that given any prime number p and a positive integer r, there is a finite field with $q = p^r$ elements. For, consider the splitting field K of $f(x) = x^q - x$ over $\mathbf{F}_p$. By Exercise 6.10, the number of roots of $f(x)$ in K is q and it is not difficult to check that these roots form a subfield of K. In fact, it has to be equal to K.

Theorem 6.4. (**Warning's theorem**) Let p be a prime and $q = p^r$ for some integer $r \geq 1$. Let $f(x_1, x_2, \cdots, x_n) \in \mathbf{F}_q[x_1, x_2, \cdots, x_n]$ be of degree less than n. Then the number of solutions of the equation $f(x_1, x_2, \cdots, x_n) = 0$ in $\mathbf{F}_q^n = \underbrace{\mathbf{F}_q \times \cdots \times \mathbf{F}_q}_{n \text{ times}}$ is divisible by p.

Proof: Let $g(x_1, x_2, \cdots, x_n) = 1 - (f(x_1, x_2, \cdots, x_n))^{q-1}$. Then the degree of g is less than $n(q-1)$. For $(\alpha_1, \alpha_2, \cdots, \alpha_n) \in \mathbf{F}_q^n$, $f(\alpha_1, \alpha_2, \cdots, \alpha_n) = 0$ if and only if $g(\alpha_1, \alpha_2, \cdots, \alpha_n) = 1$.

Therefore, the number of solutions of the equation $f(x_1, x_2, \cdots, x_n) = 0$ in $\mathbf{F}_q^n$ is $\sum g(\alpha_1, \alpha_2, \cdots, \alpha_n)$ where the sum is over all $(\alpha_1, \alpha_2, \cdots, \alpha_n) \in \mathbf{F}_q^n$.

We claim that this sum is $0 \in \mathbf{F}_q$. Once this claim is established, we shall be through. If g had been a monomial $x_1^{i_1} \cdots x_n^{i_n}$, then the sum would be $(\sum_{\alpha_1 \in \mathbf{F}_q} \alpha_1^{i_1}) \cdots (\sum_{\alpha_n \in \mathbf{F}_q} \alpha_1^{i_n})$. At least one of the i_j's (say i_l) is less than $q-1$ because of the degree consideration. Now, there is $a \in \mathbf{F}_q^*$ such that $a^{i_l} \neq 1$. Now, $\sum_{\alpha_l \in \mathbf{F}_q} \alpha_l^{i_l} = \sum_{\alpha_l \in \mathbf{F}_q} a^{i_l} \alpha_l^{i_l}$ and since $a^{i_l} \neq 1$, $\sum_{\alpha_l \in \mathbf{F}_q} \alpha_l^{i_l} = 0$. Therefore the product $(\sum_{\alpha_1 \in \mathbf{F}_q} \alpha_1^{i_1}) \cdots (\sum_{\alpha_n \in \mathbf{F}_q} \alpha_1^{i_n}) = 0$.

The general case follows trivially as that would be sum of constant multiples of these sums. □

Theorem 6.5. (**Chevalley's Theorem**)

Let $f(x_1, x_2, \cdots, x_n) \in \mathbf{F}_q[x_1, x_2, \cdots, x_n]$ be of degree less than n satisfying $f(0, 0, \cdots, 0) = 0$. Then there exists $(\alpha_1, \alpha_2, \cdots, \alpha_n) \in \mathbf{F}_q^n$ with not all α_i's zero, such that $f(\alpha_1, \alpha_2, \cdots, \alpha_n) = 0$.

Proof: Since $f(0, 0, \cdots, 0) = 0$, there is at least one solution. Now, by Theorem 6.4, the number of solutions of the equation $f(x_1, x_2, \cdots, x_n) = 0$ in $\mathbf{F}_q^n$ is a multiple of p and hence at least $p \geq 2$. This proves the theorem.

□

Remark 6.5. One can modify the construction of the polynomial g in Theorem 6.4, to get the same result as in Theorem 6.5, when one considers simultaneous solutions for any finite number of polynomials, if the sum of their degrees is less than n.

Exercise 6.11. Let p be a prime. From the above remark, deduce that given a sequence $a_1, a_2, \cdots, a_{2p-1}$ of not necessarily distinct integers, there is a subsequence of p elements whose sum is a mutiple of p.

Remark 6.6. The above exercise is true even if p is replaced by any positive integer n. Then the assertion is known as 'Erdös-Ginzburg-Ziv' theorem. To know more on this type of additive number theory one may look into [Nat 1996].

Chapter 7

Quadratic reciprocity law

Let $f(x)$ be a polynomial with integer coefficients. The problem of determining solutions to the polynomial congruence $f(x) \equiv 0 \pmod{n}$ boils down to (exercise 1.8) solving the congruence modulo prime powers dividing n. In fact, the problem can be further reduced to solving congruences modulo primes together with a set of linear congruences. To find a general method for solving polynomial congruences modulo a prime is one of the most important unsolved problems in the theory of numbers. The first nontrivial case is the quadratic congruence

$$ax^2 + bx + c \equiv 0 \pmod{p}, \ a, b, c \in \mathbf{Z} \text{ and } \ a \not\equiv 0 \bmod p.$$

Completing square, it reduces to solving a congruence of the type

$$x^2 \equiv d \pmod{p}, \ d \in \mathbf{Z}, \quad p \text{ a rational prime.} \tag{7.1}$$

7.1 Law of quadratic reciprocity

In this section, we give an algebraic treatment of the the law of quadratic reciprocity, one of the most celebrated results in all of number theory, which addresses to the question of existence of solutions to the congruence (7.1). A sketch of an elementary proof is given in an exercise at the end of this chapter.

Let $q = p^n$ and let $\mathbf{F}_q$ be the field with q elements. If $p = 2$, then since $x \mapsto x^2$ is an automorphism of $\mathbf{F}_q$, we conclude that all elements of $\mathbf{F}_q$ are squares.

If $p \neq 2$, then let Ω be an algebraic closure of $\mathbf{F}_q$ and for $x \in \mathbf{F}_q^*$ let $y \in \Omega$ be such that $y^2 = x$.

Then, $y^{q-1} = x^{(q-1)/2} = \pm 1$, since $x^{q-1} = 1$. For x to be a square in $\mathbf{F}_q$, it is necessary and sufficient that y belongs to $\mathbf{F}_q^*$, i.e., $y^{q-1} = 1$. So, if we consider the map $x \mapsto x^{(q-1)/2}$ from $\mathbf{F}_q^*$ to $\{\pm 1\}$, then this map is clearly a homomorphism and $\mathbf{F}_q^{*2}$ is the kernel. There are several ways to see that this homomorphism is surjective. One way is by observing that for an element a in $\mathbf{F}_q^*$, both a and $-a$ (they are different as the characteristic is not 2) have the same square and then using the counting argument. One may also observe that $\mathbf{F}_q^*$ is a cyclic group of even order. Therefore, it follows that the index of $\mathbf{F}_q^{*2}$ is 2.

Definition. For a prime p, other than 2, and for $x \in \mathbf{F}_q^*$, we define the *Legendre symbol* $\left(\frac{x}{p}\right)$ as $x^{(p-1)/2}$.

We extend $\left(\frac{x}{p}\right)$ to all of $\mathbf{F}_p$ by putting $\left(\frac{0}{p}\right) = 0$ and can view $\left(\frac{x}{p}\right)$ as a function on $\mathbf{Z}$ in the obvious way.

For $x \not\equiv 0 \pmod p$, $\left(\frac{x}{p}\right) = 1$ or -1 according as x is a square mod p or not, i.e., $y^2 \equiv x \pmod p$ has a solution or not. One says that x is *quadratic residue* or *quadratic non-residue* mod p respectively.

For a rational prime p, other than 2, since the index of $\mathbf{F}_p^{*2}$ in $\mathbf{F}_p^*$ is 2, there are as many residues as non-residues mod p. We also know that $\left(\frac{x}{p}\right) \cdot \left(\frac{y}{p}\right) = \left(\frac{xy}{p}\right)$, i.e., the Legendre symbol is a homomorphism from $\mathbf{F}_q^*$ to the mutiplicative group $\mathbf{C}^*$ of non-zero complex numbers (a homomorphism from an abelian group to $\mathbf{C}^*$ is called a *character* of the group).

Theorem 7.1. (Laws of quadratic reciprocity). If p and l are two odd primes,

i) $\left(\frac{-1}{p}\right) = (-1)^{\frac{p-1}{2}}$.

ii) $\left(\frac{2}{p}\right) = (-1)^{\frac{p^2-1}{8}}$.

iii) $\left(\frac{p}{l}\right)\left(\frac{l}{p}\right) = (-1)^{\frac{p-1}{2}\cdot\frac{l-1}{2}}$.

Proof: Part i) directly follows from the definition. We remark that in Exercise 1.6 and Theorem 3.1 (iii), this result had been already seen.

Now, let α denote a primitive 8th root of unity in an algebraic closure Ω of $\mathbf{F}_p$. (Since $p^{\phi(8)} \equiv 1 \pmod 8$, if we take a finite field F of dimension $\phi(8)$ over $\mathbf{F}_p$, then F itself will contain a primitive 8th root of unity. By the same reasoning, Ω contains primitive sth root of unity for any positive integer s with $(s,p) = 1$.)

Then,

$$\alpha^4 = -1. \tag{7.2}$$

Therefore we have, $(\alpha^2 + \alpha^{-2})^2 = 0$, which implies

$$\alpha^2 + \alpha^{-2} = 0. \tag{7.3}$$

Writing,

$$y = \alpha + \alpha^{-1}, \tag{7.4}$$

by (7.3), we have

$$y^2 = 2. \tag{7.5}$$

From (7.4), it follows that $y^p = \alpha^p + \alpha^{-p}$.

Therefore, in the case $p = \pm 1 \pmod 8$, we have $y^p = y$ (since $\alpha^8 = 1$), which implies $y^{p-1} = 1$ and hence

$$\left(\frac{2}{p}\right) = 2^{\frac{p-1}{2}} = y^{p-1} = 1. \quad (\text{ by } (7.5))$$

In the case $p = \pm 5 \pmod 8$, we have $y^p = -y$. Therefore, $y^{p-1} = -1$ which gives $\left(\frac{2}{p}\right) = -1$. This proves (ii).

To prove (iii), let ω be a primitive l-th root of unity in Ω.

We form the Gauss-sum $S = \sum_{x \in \mathbf{F}_l^*} \left(\frac{x}{l}\right) \omega^x$. Note that ω^x is well defined for $x \in \mathbf{F}_l$.

$$\begin{aligned} \text{We have, } S^2 &= \sum_{x,y \in \mathbf{F}_l^*} \left(\frac{xy}{l}\right) \omega^{x+y} \\ &= \sum_{y,z \in \mathbf{F}_l^*} \left(\frac{y^2 z}{l}\right) \omega^{y(z+1)} \quad (\text{Putting } x = yz) \\ &= \sum_{y,z \in \mathbf{F}_l^*} \left(\frac{z}{l}\right) \omega^{y(z+1)} \end{aligned}$$

$$\begin{aligned}
&= \sum_{y\in\mathbf{F}_l^*}\left(\frac{-1}{l}\right)\omega^0 + \sum_{z\neq -1}\left(\frac{z}{l}\right)\sum_{y\in\mathbf{F}_l^*}\omega^{y(z+1)} \\
&= \left(\frac{-1}{l}\right)(l-1) + (-1)\sum_{z\neq -1}\left(\frac{z}{l}\right)
\end{aligned}$$

(Since, $\sum_{y\in\mathbf{F}_l^*}\omega^{y(z+1)} + 1 = 1+\omega+\cdots+\omega^{l-1} = 0$),

$$S^2 = l\left(\frac{-1}{l}\right) - \sum_{z\in\mathbf{F}_l^*}\left(\frac{z}{l}\right).$$

Now, there are as many squares as non-squares in $\mathbf{F}_l^*$, $\sum_{z\in\mathbf{F}_l^*}\left(\frac{z}{l}\right) = 0$ and hence

$$S^2 = l\left(\frac{-1}{l}\right). \tag{7.6}$$

Since Ω is of characteristic p, we have

$$\begin{aligned}
S^p &= \sum_{x\in\mathbf{F}_l^*}\left(\frac{x}{l}\right)\omega^{xp} \\
&= \sum_{z\in\mathbf{F}_l^*}\left(\frac{zp^{-1}}{l}\right)\omega^{z} \quad \left(\begin{array}{c}\text{writing } p^{-1} \text{ for the inverse of } p \text{ in } F_l^* \\ \text{and putting } xp = z\end{array}\right) \\
&= \left(\frac{p^{-1}}{l}\right)S \\
&= \left(\frac{p}{l}\right)S.
\end{aligned}$$

It is clear from (7.6) that $S \neq 0$ and therefore

$$S^{p-1} = \left(\frac{p}{l}\right). \tag{7.7}$$

From (7.6) and (7.7),

$$\left(\frac{p}{l}\right) = S^{p-1} = \left(l\left(\frac{-1}{l}\right)\right)^{\frac{p-1}{2}} = \left(\frac{l}{p}\right)\left(\frac{-1}{l}\right)^{\frac{p-1}{2}} = \left(\frac{l}{p}\right)(-1)^{\frac{p-1}{2}\cdot\frac{l-1}{2}}$$

This proves the equality in (iii) modulo p. Since p is odd, the result follows. □

Remark 7.1. In Theorem 7.1, actually iii) is the proper reciprocity law while i) and ii) are respectively the first and the second supplimentary laws.

Corollary 7.1.1. Any quadratic extension K of $\mathbf{Q}$ is contained in $\mathbf{Q}(\zeta)$ for a root of unity ζ.

Proof: In the proof of iii) above, if we replace ω by a primitive l-th root of unity in $\overline{\mathbf{Q}}$, then defining S in the same way, we have $S^2 = \pm l$. Thus square root of any odd prime is contained in $\mathbf{Q}(\zeta)$ for some root of unity ζ. Observing that $\sqrt{2} \in \mathbf{Q}(\zeta_8)$ where ζ_8 is a primitive eighth root of unity in $\bar{\mathbf{Q}}$ (for $2 = -i(1+i)^2$), the result follows. □

***Remark 7.2.** The above corollary is a special case of a famous result conjectured by Kronecker and proved by Weber. We here just state the result:

Any abelian extension K of $\mathbf{Q}$ (that is, a Galois extension K of $\mathbf{Q}$ such that the Galois group Gal $(K/\mathbf{Q})$ is abelian) is contained in $\mathbf{Q}(\zeta)$ for a root of unity ζ.

When the base field $\mathbf{Q}$ is replaced by an imaginary quadratic field $K = Q(\sqrt{d}), d < 0$, then the role of ζ is taken by the coordinates of points of finite order on certain elliptic curves.

Exercise 7.1. Decide whether 45 is a square modulo the prime 1009.

Exercise 7.2. Does the Diophantine equation $y^2 = x^3 + 23$ have solutions?

Remark 7.3. The equation considered in the above exercise is a special case of Bachet equation $y^2 = x^3 + k$, which was discussed by Bachet in 1621. It was proved by Mordell that such an equation has only a finite number of solutions.

Exercise 7.3. Show that there are infinitely many primes of the form $8n - 1$.

7.2 The Jacobi symbol

Definition. Given an integer a and an odd positive integer b, we define the Jacobi symbol $\left(\frac{a}{b}\right)$ which extends the Legendre symbol.

Let the prime factorization of the odd positive integer b be given by $b = \prod_{i=1}^{r} p_i^{n_i}$. Then the *Jacobi symbol* is defined by

$$\left(\frac{a}{b}\right) = \prod_{i=1}^{r} \left(\frac{a}{p_i}\right)^{n_i}$$

Remark 7.4. If a is a square modulo the integer b, that is, if the congruence $x^2 \equiv a \pmod{b}$ has a solution, then since each $\left(\frac{a}{p_i}\right) = 1$, it follows that $\left(\frac{a}{b}\right) = 1$. However, the converse is not true.

Exercise 7.4. For integers a, a', b and b', where b and b' are odd positive integers, one has

1) $\left(\frac{a}{b}\right)\left(\frac{a'}{b}\right) = \left(\frac{aa'}{b}\right)$.

2) $\left(\frac{a}{b}\right)\left(\frac{a}{b'}\right) = \left(\frac{a}{bb'}\right)$.

3) $\left(\frac{a}{b}\right) = \left(\frac{a'}{b}\right)$, whenever $a \equiv a' \pmod{b}$.

Exercise 7.5. Let a and b be odd positive integers. Then,

1) $\left(\frac{-1}{b}\right) = (-1)^{\frac{b-1}{2}}$.

ii) $\left(\frac{2}{b}\right) = (-1)^{\frac{b^2-1}{8}}$.

iii) $\left(\frac{a}{b}\right)\left(\frac{b}{a}\right) = (-1)^{\frac{a-1}{2}\cdot\frac{b-1}{2}}$.

7.3 Applications

First, we see an interesting application of part (i) of Theorem 7.1 in additive number theory.

Reading modulo 4, it can be easily seen that an integer $n \equiv 3 \bmod 4$ can not be expressed as a sum of two squares in $\mathbf{Z}$.

On the other hand, we now prove:

Theorem 7.2. Every prime p congruent to 1 mod 4 can be written as a sum of two squares in $\mathbf{Z}$.

Proof: Let p be a prime congruent to 1 mod 4. First we show that if $\exists x, y \in \mathbf{Z}$ such that $x^2 + y^2 = kp$ for an integer $k \geq 2$, $k \neq p$ then $\exists x_1, y_1 \in \mathbf{Z}$ such that $x_1^2 + y_1^2 = k_1 p$ for $1 \leq k_1 < k$.

We choose

$$x_0 \equiv x \pmod{k}, y_0 \equiv y \pmod{k}, \text{ where } -\frac{k}{2} \leq x_0, y_0 < \frac{k}{2}$$

Then, $x_0^2 + y_0^2 \equiv x^2 + y^2 \equiv 0 \pmod{k}$. Let $x_0^2 + y_0^2 = k_1 k$.

Now, $$(x_0 x + y_0 y)^2 + (x_0 y - y_0 x)^2 = (x_0^2 + y_0^2)(x^2 + y^2) = k_1 k^2 p. \quad (7.8)$$

But,

$$x_0 x + y_0 y \equiv x^2 + y^2 \equiv 0 \pmod{k}$$

and

$$x_0 y - y_0 x \equiv x y_0 - y_0 x \equiv 0 \pmod{k}$$

and hence from (7.8),

$$\left(\frac{x_0 x + y_0 y}{k}\right)^2 + \left(\frac{x_0 y - y_0 x}{k}\right)^2 = k_1 p.$$

giving us two integers x_1, y_1 satisfying $x_1^2 + y_1^2 = k_1 p$.

Since, $k_1 k = x_0^2 + y_0^2 \leq \frac{k^2}{2}$ we have $k_1 \leq \frac{k}{2}$. Also, $k_1 = 0$ gives $x_0 = 0, y_0 = 0$. This implies $x \equiv 0 (\text{mod } k), y \equiv 0 (\text{mod } k)$.

Hence, $k^2 \mid (x^2 + y^2) = pk$, which gives $k = p$ and this is a contradiction. Therefore, $1 \leq k_1 < k$.

Since -1 is a quadratic residue mod p, there is an integer u, $2 \leq u \leq p-1$ such that $u^2 + 1 = kp$ for some $k \geq 1$. Also, since the difference of two squares can not be 1, we have $k \neq p$. If $k = 1$, we are through. Otherwise by the above result we get $x_1, y_1 \in \mathbf{Z}$ such that $x_1^2 + y_1^2 = k_1 p$ for $1 \leq k_1 < k$ and continuing this way we get integers r and s such that $r^2 + s^2 = p$. □

Theorem 7.3. An integer $n \geq 1$ is a sum of two squares if and only if no prime $p \equiv 3 \pmod{4}$ occurs to an odd power in the factorization of n into powers of distinct primes.

Proof: First, suppose that $n = x^2 + y^2$ and let $p \equiv 3 \pmod{4}$ be a prime dividing n. Let p^r be the highest power of p dividing n.

If possible let r be odd. If $d = (x, y)$, the gcd of x and y, then $d^2 \mid n$ and

$$n_1 = x_1^2 + y_1^2, \quad (7.9)$$

where $x_1 = x/d, y_1 = y/d$ and $n_1 = n/d^2$. Now, $(x_1, y_1) = 1$ and therefore p can divide at most one of the integers x_1, y_1 and n_1. Since r is odd, $p \mid n_1$. Therefore p does not divide either of x_1, y_1. Regarding (7.9) as an equation over F_p, we now have $-1 = (x_1/y_1)^2$ which means

$$\left(\frac{-1}{p}\right) = 1.$$

But this is impossible, because $p \equiv 3 \pmod 4$, which implies that

$$\left(\frac{-1}{p}\right) = (-1)^{\frac{p-1}{2}} = -1.$$

Thus r cannot be odd.

Conversely, if no prime $p \equiv 3 \pmod 4$ occures to an odd power in the factorization of n into powers of distinct primes, then $n = m^2 p_1 \cdots p_l$, where $p_j \equiv 1 \pmod 4, j = 1, \cdots, l$. Since from (7.8), product of two sums of squares is again a sum of squares, by Theorem 7.2, n is a sum of two squares. □

Suppose we are given an integer n, which is a square. Then, of course, n is a quadratic residue mod p for every prime p. One may ask whether the converse is true. That is, whether for an integer n, being a quadratic residue mod p for every prime p does necessarily imply that n is a square.

The following theorem asserts something stronger than that. Our proof closely follows the one given in [IR].

Theorem 7.4. If an integer is a quadratic residue modulo all but finitely many primes, then it is a square.

Proof: We prove that if a positive integer a is not a square, then there are infinitely many primes p such that a is a quadratic non-residue mod p. Since for an odd prime p, we have $\left(\frac{-1}{p}\right) = (-1)^{\frac{p-1}{2}}$, this will give the result.

Since a positive non-square a can be written as $n^2 a'$, where $a' > 1$ is square-free, we can assume a to be square-free to start with. Therefore, let $a > 1$ be square-free. Let $a = 2^s p_1 p_2 \cdots p_r$ where $s = 0$, or 1 and p_i's are distinct odd primes.

The proof is divided into two cases, the first one where $r = 0$ (and hence necessarily $s = 1$) and the second one where $r > 0$.

Case I $(r = 0,\ s = 1)$

Here $a = 2$. Let $5 = q_1, q_2, \cdots, q_m$ be a finite set of odd primes not including 3 such that $\left(\frac{2}{q_i}\right) = -1$, for all $i = 1, 2, \cdots m$.

Let $b = 8q_1q_2\cdots q_m + 3$. Then, by Exercise 7.5 (2), we have $\left(\frac{2}{b}\right) = -1$ and hence $\left(\frac{2}{l}\right) = -1$ for some prime divisor l of b. But l can neither be 3 nor any of the primes q_i. Hence there are infinitely many odd primes modulo which 2 is a quadratic non-residue.

Case II $(r > 0)$

Let $q_1, q_2, \cdots, q_m$ be a finite set of odd primes not including any of the p_i's. Let t be an integer satisfying $\left(\frac{t}{p_r}\right) = -1$.

By Chinese remainder theorem, there is a positive integer N satisfying the set of following congruences:

$$\begin{aligned} x &\equiv 1 \pmod{q_i}, && \text{for } i = 1, 2, \cdots, m \\ x &\equiv 1 \pmod{8}, && \\ x &\equiv 1 \pmod{p_i}, && \text{for } i = 1, 2, \cdots, r-1 \\ x &\equiv t \pmod{p_r}. && \end{aligned}$$

Since $N \equiv 1 \pmod 8$, from Exercise 7.4, we have $\left(\frac{2}{N}\right) = 1$ and $\left(\frac{p_i}{N}\right) = \left(\frac{N}{p_i}\right)$, for $i = 1, 2, \cdots, r$.

Therefore,

$$\begin{aligned} \left(\frac{a}{N}\right) &= \left(\frac{2}{N}\right)\left(\frac{p_1}{N}\right)\cdots\left(\frac{p_{r-1}}{N}\right)\left(\frac{p_r}{N}\right) \\ &= \left(\frac{2}{N}\right)\left(\frac{N}{p_1}\right)\cdots\left(\frac{N}{p_{r-1}}\right)\left(\frac{N}{p_r}\right) \\ &= -1 \end{aligned}$$

Hence from the definition of the Jacobi symbol, it follows that $\left(\frac{a}{p}\right) = -1$ for some prime p dividing N. Also, p is an odd prime such that $p \notin \{q_1, q_2, \cdots, q_m\}$. □

7.4 An elemetary approach

In the following exercise, we sketch an elementary proof of the law of quadratic reciprocity.

Exercise 7.6. Let p be an odd prime and a be an integer such that $a \not\equiv 0 \pmod{p}$. We consider the least positive residues, t_i, $i = 1, 2, \cdots, (p-1)/2$, $\pmod{p}$ of the following $(p-1)/2$ multiples of a:

$$a, 2a, \cdots, \frac{p-1}{2}a. \tag{7.10}$$

i) Observe that the numbers in (7.10) are incongruent modulo p.

ii) Let $r_1, r_2, \cdots, r_m$ denote those among these least residues modulo p which are less than or equal to $(p-1)/2$ and $s_1, s_2, \cdots, s_n$ denote those that are greater than $(p-1)/2$.

Then $m + n = (p-1)/2$ and show that the $(p-1)/2$ integers

$$r_1, r_2, \cdots, r_m, p - s_1, \cdots, p - s_n$$

are all distinct.

Hence deduce (**Gauss**):

$$\left(\frac{a}{p}\right) = (-1)^n.$$

iii) Using above result of Gauss, prove that:

$$\left(\frac{2}{p}\right) = (-1)^{\frac{p^2-1}{8}}$$

iv) If $k \in \{1, 2, \cdots, \frac{p-1}{2}\}$, we have $ka = [ka/p]p + t_k$, where $[ka/p]$ denotes the integral part of (ka/p). Consider the sums

$$\sum_{k=1}^{(p-1)/2} ka \text{ and } \sum_{k=1}^{(p-1)/2} k = \sum_{k=1}^{m} r_k + \sum_{k=1}^{n} (p - s_k).$$

Subtracting the second sum from the first, prove that if a happens to be odd, then

$$\left(\frac{a}{p}\right) = (-1)^M, \text{ where } M = \sum_{k=1}^{(p-1)/2} [ka/p].$$

v) Let p and q be two odd primes. Consider the rectangle on the XY-plane having vertices at $(0,0), (p/2,0), (0,q/2)$ and $(p/2,q/2)$. Observing that there are $[kq/p]$ integer lattice points above the point $(k,0)$ on the vertical line segment joining the points $(k,0)$ and $(k,kq/p)$ and making a similar observation for horizontal line segments, show that the counting of integer lattice points in the interior of the rectangle leads to a proof of part (iii) of Theorem 7.1.

Exercise set B

B.1 Let R be a Euclidean domain. Show that there is a non-zero non-unit u in R such that for any $\alpha \in R$, $\exists r \in R^* \cup \{0\}$ with $u|(\alpha - r)$.

B.2 Show that the ring $R = \mathbf{Z} + \mathbf{Z}\frac{1+\sqrt{-13}}{2}$, is not a Euclidean domain.

B.3 For a prime integer p, prove that the polynomial $f(x) = x^{p-1} + x^{p-2} + \cdots + 1$ is irreducible in $\mathbf{Q}\,[x]$.

B.4 Prove that the polynomials $x^2 + 1$ and $x^2 + x + 4$ are irreducible elements in the ring $\mathbf{F}_{11}[x]$. Show that the fields $\frac{\mathbf{F}_{11}[x]}{(x^2+1)}$ and $\frac{\mathbf{F}_{11}[x]}{(x^2+x+4)}$ are isomorphic.

B.5 Let $K = \mathbf{F}_2/(1 + x + x^2)$. What is Gal $(K : \mathbf{F}_2)$?

B.6 Let $q = p^r$ where p is a prime. Show that Gal$(\mathbf{F}_q/\mathbf{F}_p)$ is the cyclic group of order n, the generator being the Frobenius automorphism $a \mapsto a^p$.

B.7 Given any mapping $f : \mathbf{F}_q \to \mathbf{F}_q$ on a finite field to itself, show that there exists an element $g(x) \in \mathbf{F}_q[x]$ such that $f(x) = g(x)\ \forall\ x \in \mathbf{F}_q$.

Conversely, if R is a ring (commutative as usual) such that every map $R \to R$ is given by a polynomial in $R[x]$, then R is a finite field.

B.8 Are there integers x, y such that $2x^2 + 3x + 1 = 7y$?

B.9 Show that the equation $y^2 = x^3 + 45$ has no solutions in integers.

Chapter 8

Modules

For a ring R, the term R-module has been already defined in Chapter 0. In this chapter, we shall see some definitions and results concerning modules. We shall restrict ourselves to gather just that amount of information as will be sufficient for our requirements while studying number fields in the remaining chapters.

8.1 Basic definitions

Definition. Let M be an R-module. A *submodule* M' of M is a subgroup of M which is closed under scalar multiplication by elements of R.

If M' is a submodule of M, then the quotient group M/M' becomes an R-module under the scalar multiplication defined by $r(a + M') = ra + M'$. This module is called the *quotient module* of M by M'.

Let M and N be two R-modules. A map $f : M \to N$ is said to be an *R-module homomorphism* if it satisfies the following conditions:

i) $f(x + y) = f(x) + f(y)$, for all $x, y \in M$,

ii) $f(rx) = rf(x)$, for all $r \in R$ and $x \in M$.

A bijective module homomorphism is called a *module isomorphism.* As in the cases of groups and rings, if $f : M \to N$ is an R-module homomorphism, then the kernel $\ker(f) = \{x \in M : f(x) = 0\}$ is a submodule of

M and the quotient module $M/\ker(f)$ is isomorphic to the image $f(M)$ (written as $\operatorname{Im}(f)$) of f.

For an element $r \in R$ and an R-module M, rM is defined to be the set $\{rm : m \in M\}$. For an ideal I of R, IM is defined to be the subset of M consisting of all finite sums $\sum r_i a_i$, $r_i \in I, a_i \in M$. In particular, this defines the product of two ideals of R. It is easy to see that rM and IM as defined above are submodules of M.

If N and N' are two submodules of an R-module M, then the set $\{r \in R : rN' \subset N\}$ is clearly an ideal of R. It is denoted by $(N : N')$. In the particular case, when $N = \{0\}$, the ideal $(\{0\} : N')$ is called the *annihilator* of N' and is denoted by $\operatorname{Ann}_R(N')$. If the annihilator $\operatorname{Ann}_R(N')$ of N' is the zero ideal, then N' is said to be *faithful.*

Let $\{M_i\}_{i\in I}$ be a family of R-modules. Then the *direct product* $\prod_{i\in I} M_i$ is the set of families $(a_i)_{i\in I}$, with $a_i \in M_i$, indexed by I where addition and scalar multiplication are defined in the obvious componentwise way. The *direct sum* $\bigoplus_{i\in I} M_i$ of these modules is defined to be the submodule of the direct product consisting of those elements $(a_i)_{i\in I}$ such that $a_i = 0$ for all but finitely many indices $i \in I$. If the index set I is finite, then direct sum and direct product defined above are the same.

A direct sum of copies of the R-module R is called a *free* R-module. We shall be using the notation R^n for the direct sum of n copies of R. By convention, R^0 denotes the zero module $\{0\}$.

If M is an R-module and S a subset of M, then the smallest submodule of M containing the set S is called the *submodule generated* by S. It is the intersection of all submodules of M containing S. Furthermore, the submodule generated by S can be explicitly described. It consists of all elements of the form $\sum_i r_i s_i$, where the sum involved is finite and $s_i \in S$, $r_i \in R$ for all i.

A submodule of M which is generated by a finite set is called *finitely generated.*

Given a family $\{M_i\}_{i\in I}$ of submodules of an R-module M, the smallest submodule containing the M_i's is called the sum of the submodules M_i and is denoted by $\sum_{i\in I} M_i$. When I is finite, say given by $I = \{1, 2, \cdots, r\}$, we shall sometimes write $M_1 + M_2 + \cdots + M_r$ for $\sum_{i=1}^{r} M_i$. The submodule generated by a subset S of M is nothing but the sum of the submodules Ra, $a \in S$.

A subset S of an R-module M is said to be *linearly independent* if for any finite subset $\{a_1, \cdots, a_t\}$ of S, $\sum_{i=1}^{t} r_i a_i = 0$, $r_i \in R$ implies $r_i = 0$ for all i. If there is a linearly independent subset S of M such that M is generated by S, then S is said to be a *basis* of M.

Remark 8.1. A vector space is a free module over a field. But unlike in the case of a vector space, a linearly independent subset of a free module need not necessarily be extendable to a basis. Also, if a subset S generates a free module, S need not necessarily contain a basis.

Exercise 8.1. Give an example of a linearly independent subset of some free module which can not be extended to a basis. Also, give an example of a subset S of some free module M such that S generates M, but S does not contain a basis of M.

Exercise 8.2.

(a) An R module M is free if and only if it has a basis.

(b) For a free R-module M, the cardinalities of any two bases of M over R are equal.

Definition. In Exercise 8.1 (b), the common cardinalities of different bases of M is called the *rank* of the free module M. We shall denote it by $\text{rank}_R(M)$.

8.2 A result on finitely generated modules

Theorem 8.1. Let M be a finitely generated R-module generated by n elements. Let $\phi : M \to M$ be an R-module homomorphism and I an ideal of R such that $\phi(M) \subset IM$. Then there is a relation of the form

$$\phi^n + a_1\phi^{n-1} + \cdots + a_n = 0.$$

Proof: Let $M = \sum_{i=1}^{n} R\omega_i$. By our assumption, there are $a_{ij} \in I$ such that $\phi(\omega_i) = \sum_{j=1}^{n} a_{ij}\omega_j$ for $1 \leq i \leq n$.

We can rewrite this as

$$\sum_{j=1}^{n}(\phi\delta_{ij} - a_{ij})\omega_j = 0 \quad \text{for } 1 \le i \le n,$$

where δ_{ij} is the Kronecker symbol.

By multiplying the left-hand side by the adjoint of the matrix $(\delta_{ij}\phi - a_{ij})$, we see that the determinant is the zero homomorphism since $\det(\delta_{ij}\phi - a_{ij})\omega_j = 0$ for all j. Expanding the determinant, we get our result. □

The following result attributed to Krull, Azumaya and Nakayama is known as *Nakayama's lemma* in literature.

Theorem 8.2. Let M be a finitely generated R-module and I an ideal of R. If $IM = M$, then there exists $r \in R$ such that $r \equiv 1 \pmod{I}$ (that is, $r - 1 \in I$) and $rM = 0$. If in addition, I is contained in the intersection (called the *Jacobson radical* of R) of all maximal ideals of R, then $M = 0$.

Proof: Taking $\phi : M \to M$ to be the identity map, from Theorem 8.1 we see that $r = 1 + a_1 + \cdots + a_n = 0$ satisfies our requirements.

Now, let I be contained in the Jacobson radical. Then r is a unit in R. For, otherwise r will be contained in some maximal ideal and from the above relation, 1 will belong to that maximal ideal, which is absurd. Therefore, $M = r^{-1}rM = 0$. □

8.3 Noetherian modules

Definition. Let M be an R-module. Then M is said to be *Noetherian* if every ascending chain $M_1 \subset M_2 \subset \cdots$ of submodules becomes stationary, that is, there is a positive integer n such that $M_n = M_{n+1} = \cdots$.

A ring R is said to be a *Noetherian ring* if it is Noetherian as an R-module.

Examples. If M is a module consisting of only finitely many elements, then it is clearly Noetherian. In particular, a finite abelian group considered as a $\mathbf{Z}$-module is Noetherian. Also, since the only ideals of $\mathbf{Z}$ are of the form $m\mathbf{Z}$ for some integer m, it is easy to see that $\mathbf{Z}$ is a Noetherian

Z-module. Thus **Z** is an example of a Noetherian ring. The polynomial ring $R[x_1, x_2, \cdots]$ in an infinite number of variables is not Noetherian.

Theorem 8.3. Let M be an R-module. Then the following conditions are equivalent:

i) M is Noetherian.

ii) Every non-empty set of submodules has a maximal element.

iii) Every submodule of M is finitely generated.

Proof: i) $\Longrightarrow$ ii)

If possible, suppose there is a non-empty set of submodules $\mathcal{M}$ which does not have a maximal element. Let $M_1 \in \mathcal{M}$. Since M_1 is not maximal, there is a submodule M_2 such that $M_1 < M_2$. Since M_2 is not maximal, there is a submodule M_3 such that $M_2 < M_3$. Proceeding in this way, we obtain an ascending chain of submodules which is not stationary, thereby contradicting the hypothesis that M is Noetherian.

ii) $\Longrightarrow$ iii)

Let N be any submodule of M. Let $\mathcal{S}$ be the set of all finitely generated submodules of N. Since $\{0\} \in \mathcal{S}$, we see that $\mathcal{S}$ is non-empty. Therefore, $\mathcal{S}$ has a maximal element N' say. Let a be any element of N. We consider the submodule $N' + Ra$ of N. This is finitely generated. Maximality of N' implies that $N' + Ra = N'$. This implies that $a \in N'$. Since $a \in N$ was arbitrary, we get $N = N'$. Hence N is finitely generated.

iii) $\Longrightarrow$ i)

Let $M_1 \subset M_2 \subset \cdots$ be an ascending chain of submodules. Now, $\cup_{r=1}^{\infty} M_i$ is again a submodule of M and by our assumption, it is finitely generated. Let $a_1, \cdots, a_m$ be a set of generators of M. Let $a_i \in M_{t_i}$. Let T be the largest among the integers $t_1, t_2, \cdots, t_m$. Then, $\cup_{r=1}^{\infty} M_i = \cup_{r=1}^{T} M_i = M_T$ and hence $M_T = M_{T+1} = M_{T+2} = \cdots$. □

Remark 8.2. We have already seen that the ring **Z** is Noetherian. From (iii) above, it follows that every principal ideal domain is Noetherian. In particular, every Euclidean domain is Noetherian.

Definition. A sequence

$$\cdots \longrightarrow M_{r-1} \xrightarrow{f_r} M_r \xrightarrow{f_{r+1}} M_{r+1} \longrightarrow \cdots$$

of R-modules $\{M_i\}$ and R-module homomorphisms $\{f_i\}$ is said to be *exact* at M_r if $\text{Im}(f_r) = \ker(f_{r+1})$. The sequence is called an *exact sequence* if it is exact at all M_i's.

An exact sequence of the special form

$$\{0\} \to M' \xrightarrow{f} M \xrightarrow{g} M'' \to \{0\}$$

is called a *short exact sequence.* This means that f is injective, g is surjective and $\text{Im}(f) = \ker(g)$.

Theorem 8.4. Let $\{0\} \to M' \xrightarrow{f} M \xrightarrow{g} M'' \to \{0\}$ be a short exact sequence of R-modules. Then M is Noetherian if and only if the modules M' and M'' are Noetherian.

Proof: Suppose M is Noetherian. Let $M_1' \subset M_2' \subset \cdots$ be an ascending chain of submodules of M'.

Now, $f(M_1') \subset f(M_2') \subset \cdots$ is an ascending chain of submodules of M. Since M is Noetherian by assumption, this sequence becomes stationary. Since f is injective, this implies that the original sequence $M_1' \subset M_2' \subset \cdots$ becomes stationary. Therefore, M' is Noetherian. Similarly, an ascending chain of submodules of M'' gives rise to an ascending chain of submodules in M and arguing similarly we get that M'' is Noetherian.

Now, suppose that the modules M' and M'' are Noetherian. Let $M_1 \subset M_2 \subset \cdots$ be an ascending chain of submodules of M. By our assumption the chains $f^{-1}(M_1) \subset f^{-1}(M_2) \subset \cdots$ and $g(M_1) \subset g(M_2) \subset \cdots$ becomes stationary. Therefore, there is a positive integer N such that $f^{-1}(M_r) = f^{-1}(M_N)$ and $g(M_r) = g(M_N)$ for all $r \geq N$.

For some $r \geq N$, let $a \in M_r$. Since $g(M_r) = g(M_N)$, there is $b \in M_N$ such that $g(a) = g(b)$. This implies that $a - b \in \ker(g) = \text{Im}(f)$. Let $f(c) = a - b$ for some $c \in f^{-1}(M_r) = f^{-1}(M_N)$. Therefore, $a - b = f(c) \in M_N$ which implies $a \in b + M_N = M_N$. Thus we have proved that $M_r = M_N$ for all $r \geq N$. Therefore, M is Noetherian. □

Corollary 8.4.1. Let $M_1, M_2, \cdots, M_r$ be Noetherian R-modules. Then, $\bigoplus_{i=1}^{t} M_i$ is Noetherian.

Proof: From the short exact sequence

$$\{0\} \to M_2 \to M_1 \bigoplus M_2 \to M_1 \to \{0\},$$

by the above theorem $M_1 \oplus M_2$ is Noetherian. In general, from the short exact sequence

$$\{0\} \to M_t \to \bigoplus_{i=1}^{t} M_i \to \bigoplus_{i=1}^{t-1} M_i \to \{0\}$$

we derive by induction that $\bigoplus_{i=1}^{t} M_i$ is Noetherian. □

Corollary 8.4.2. Let R be a Noetherian ring and M a finitely generated R-module. Then M is a Noetherian module.

Proof: Being finitely generated, M is a quotient of R^n for some positive integer n. By Corollary 8.4.1, R^n is Noetherian. Therefore, by Theorem 8.4, it follows that M is Noetherian. □

Exercise 8.3. If a ring R is Noetherian, show that so is the polynomial ring $R[x_1, \cdots, x_r]$. (This assertion is known as **Hilbert's Basis Theorem.**)

Exercise 8.4. Let R be a ring such that all of its prime ideals are finitely generated. Show that R is Noetherian. (This result is due to **Cohen.**)

Exercise 8.5. Let $M \xrightarrow{g} M' \to \{0\}$ be an exact sequence where M, M' are R-modules. Show that given a homomorphism $f : R \to M'$ there is a homomorphism h from R to M such that $goh = f$. Show that the above result is true if one replaces R by a free R-module P. Hence show that given a short exact sequence

$$\{0\} \to M_1 \xrightarrow{f} M \xrightarrow{g} P \to \{0\}$$

of R-modules where P is free, $M_1 \bigoplus P = M$.

8.4 Modules over a PID

We now prove an important result on modules over principal ideal domains.

Theorem 8.5. Let R be a PID and M a free R- module of $\text{rank}_R(M) = m$, where m is a positive integer. If N is a submodule of M, then N is also a free R-module and $\text{rank}_R(N) \leq m$.

Proof: Let $\{a_1, a_2, \cdots, a_m\}$ be a basis of M. For $1 \leq r \leq m$, let T_r denote the submodule of M generated by the set $\{a_1, a_2, \cdots, a_r\}$. Let $N_r = N \cap T_r$.

Now, $N_1 = \{a \in N : a = ra_1, \text{ for some } r \in R\}$. The set $\{r \in R : ra_1 \in N_1$ is an ideal of R and since R is a PID, this ideal is generated by an element, say r_1, of R. If $r_1 = 0$, then N_1 is the free module $\{0\} = R^0$ generated by the empty set. If $r_1 \neq 0$, then clearly N_1 is generated by the element $r_1 a_1$.

Let $2 \leq t \leq m$. Now, assume that for $i \leq t-1$, we have proved that N_i is free with $\text{rank}_R(N_i) \leq i$. We prove that N_t is free with $\text{rank}_R(N_t) \leq t$.

Let $I = \{r \in R : ra_t + \sum_{j=1}^{t-1} r_j a_j \in N_t \text{ for some } r_j's \in R\}$. Once again, this I is an ideal of R and so I is a principal ideal generated by an element r_t say. Clearly, $r_t = 0$ if and only if $N_{t-1} = N_t$. If $r_t = 0$, then by induction hypothesis, N_t is free with $\text{rank}_R(N_t) \leq t - 1 < t$.

Assume $r_t \neq 0$. Then, there is $a' \in N_t$ such that $a' = r_t a_t + \sum_{j=1}^{t-1} r_j a_j \in N_t$ for some r_j's $\in R$. Let a'' be any element of N_t.

Then, $a'' = \sum_{j=1}^{t} s_j a_j$ for some s_j's $\in R$ where $s_t = rr_t$ for some $r \in R$. Now, $ra' - a'' \in N_{t-1}$. It follows that $N_t = N_{t-1} \bigoplus Ra'$. Hence, N_t is free with $\text{rank}_R(N_t) \leq (t-1) + 1 = t$. □

Remark 8.3. The statement of the above theorem holds even if $\text{rank}_R(M)$ is not finite.

8.5 Some special results

Keeping in view our requirements in the later chapters, in the remaining portion of this chapter, we concern ourselves with a result on bilinear forms over a vector space followed by a result on lattices in $\mathbf{R}^n$.

Definition. Let V be a vector space over a field K. A *bilinear form* B on V is a mapping $V \times V \to K$ such that for fixed $a, b \in V$, the maps $V \to K$ defined by $x \mapsto B(x, b)$ and $y \mapsto B(a, y)$ are homomorphisms from V into K.

A bilinear form $B(x, y)$ on V is said to be *non-degenerate* if for every pair of non-zero elements $a, b \in V$, the homomorphisms $x \mapsto B(x, b)$ and $y \mapsto B(a, y)$ are non-zero.

Exercise 8.6. Let V be a vector space of dimension n over a field K

and $B(x,y)$ a non-degenerate bilinear form on V. Then for any K-basis $\omega_1, \omega_2, \cdots, \omega_n$ of V, there corresponds a basis $\omega_1', \omega_2', \cdots, \omega_n'$ such that $B(\omega_i, \omega_j') = \delta_{ij}$, $1 \leq i, j \leq n$.

Definition. A subgroup H of $\mathbf{R}^n$ is called *discrete* if for any compact subset S of $\mathbf{R}^n$, the intesection $H \cap S$ is finite.

Example. $\mathbf{Z}^n \subset \mathbf{R}^n$ is discrete.

Theorem 8.6. Let H be a discrete subgroup of $\mathbf{R}^n$. Then H is generated (as a $\mathbf{Z}$-module) by r $(\leq n)$ vectors which are linearly independent over $\mathbf{R}$.

Proof: Let $r(\leq n)$ be the largest integer such that H has r elements linearly independent over $\mathbf{R}$. Let $e_1, e_2, \cdots, e_r \in H$ be linearly independent over $\mathbf{R}$.

Let

$$P = \{x \in \mathbf{R}^n | \ x = \sum_{i=1}^{r} \alpha_i e_i, 0 \leq \alpha_i \leq 1\}$$

be the parallelotope formed by using the origin and e_i's as its vertices.

Since P is compact, $P \cap H$ is finite. Let $x \in H$. From the maximality of $(e_1, e_2, \cdots, e_r)$ we get $x = \sum_{i=1}^{r} \lambda_i e_i$, $\lambda_i \in \mathbf{R}$.

For any positive integer l, write

$$x_l = lx - \sum_{i=1}^{r} [l\lambda_i] e_i = \sum_{i=1}^{r} (l\lambda_i - [l\lambda_i]) e_i \in P \cap H.$$

Since $P \cap H$ is finite, there is $j \neq k$ such that $x_j = x_k$.

This implies, $(j-k)\lambda_i = [j\lambda_i] - [k\lambda_i]$ for $1 \leq i \leq r$ and hence, $\lambda_i \in \mathbf{Q}$ for $1 \leq i \leq r$.

Thus any element of H can be written as a linear combination of the e_i's with rational coefficients.

Also, since $x = x_1 + \sum_{i=1}^{r} [\lambda_i] e_i$ where $x_1 \in P \cap H$, as a $\mathbf{Z}$-module H is generated by $P \cap H$.

Now, each element of $P \cap H$ is a $\mathbf{Q}$ -linear combination of the e_i's.

Let $d \in \mathbf{Z} - \{0\}$ be the common denominator of these coefficients (remember that $P \cap H$ is finite).

Then, $dH \subset \sum_{i=1}^{r} \mathbf{Z} e_i$.

Therefore, by Theorem 8.5, dH is free abelian of rank $\leq r$.

Also, since $H \supset \sum_{i=1}^{r} \mathbf{Z}e_i$, rank $H \geq r$. Since, H and dH have the same rank, rank $H = r$.

Since, H contains r vectors which are linearly independent over $\mathbf{R}$, the generators of H as a free $\mathbf{Z}$-module, must be linearly independent over $\mathbf{R}$. $\square$

Definition. A discrete subgroup of rank n of $\mathbf{R}^n$ is called a *lattice* in $\mathbf{R}^n$.

Remark 8.4. By Theorem 8.6, a lattice is generated over $\mathbf{Z}$ by a basis of $\mathbf{R}^n$ over $\mathbf{R}$ which is a $\mathbf{Z}$-basis for the given lattice.

Chapter 9

Gaussian integers and the ring $\mathbf{Z}[\sqrt{-5}]$

We have seen that the ring of Gaussian integers is a Euclidean domain. In the first part of this chapter, we try to have a better understanding of this ring. In the later part, we discuss the ring $\mathbf{Z}[\sqrt{-5}]$, which fails to be a UFD. The study of these rings serve as a prelude to the study of ring of integers in number fields which will be taken up in the later chapters.

9.1 Gaussian integers

We start with the following observations. If a Gaussian integer belongs to the set of real numbers, then it is an ordinary integer. Again, an ordinary integer d divides a Gaussian integer $a + ib$ in the ring of Gaussian integers, if and only if d divides both a and b. In Chapter 4, we have already seen that the ring of Gaussian integers is a Euclidean domain (and hence a PID) with the size function σ defined on it by $\sigma(a+ib) = a^2 + b^2$. Also, for two elements $a + ib, c + id \in \mathbf{Z}[i]$, $\sigma((a+ib)(c+id)) = \sigma(a+ib)\sigma(c+id)$. This is a well-known property of absolute values of complex numbers and can be easily checked.

Now, let $\alpha = a + ib$ be a unit in $\mathbf{Z}[i]$. Then there is α' in $\mathbf{Z}[i]$ such that $\alpha\alpha' = 1$ and hence $\sigma(\alpha)\sigma(\alpha') = 1$. This implies $a^2 + b^2 = \sigma(\alpha) = 1$. Conversely, if $\sigma(a + ib) = 1$, then $a^2 + b^2 = (a + ib)(a - ib) = 1$, which implies that $a + ib$ is a unit.

Now, the only solutions of the Diophantine equation $a^2 + b^2 = 1$ are $a = \pm 1,\ b = 0$ or $a = 0,\ b = \pm 1$. Hence the units in $\mathbf{Z}[i]$ are ± 1 and $\pm i$.

Definition. A prime element of the ring of Gaussian integers will be called a *Gauss prime.*

Theorem 9.1. If p is a prime integer, then either p is a Gauss prime or else it is a product of two complex conjugate Gauss primes.

Proof: Let p be an integer prime. Since the only units in the ring of Gaussian integers are $\pm 1, \pm i$, the integer prime p has a Gauss prime divisor $\pi = a + ib$, say. Since p is real it is equal to its complex conjugate $\overline{p}$, and therefore $\overline{\pi} = a - bi$ also divides p. This implies that $\pi\overline{\pi} = a^2 + b^2$ divides p^2 in the ring of Gaussian integers. But now, $\pi\overline{\pi}$ is an integer divisor of p^2. Therefore, in the ring of Gaussian integers, either π is a proper divisor of p or π is an associate of p. In the second case, p is a Gauss prime. In the former case, $\pi\overline{\pi}$ is a proper divisor of p^2 in the ring $\mathbf{Z}$, which implies that $\pi\overline{\pi} = p$. □

Theorem 9.2. If π is a Gauss prime, then $\pi\overline{\pi}$ is a prime or square of a prime in the ring of integers.

Proof: Let π be a Gauss prime. Now, $\pi\overline{\pi} = n \in \mathbf{Z}$ and therefore, in $\mathbf{Z}[i]$, the Gauss prime π will divide one of the integer prime factors, say p, of n. Now, $\pi\overline{\pi}$ being an integer divisor of p^2, the result follows. □

Theorem 9.3. If p be a prime integer, then the following are equivalent:

(i) p is a product of two complex conjugate Gauss primes.

(ii) p is the sum of two integer squares.

(iii) $p = 2$ or $p \equiv 1 (\mathrm{mod}\ 4)$.

Proof: (i) $\Rightarrow$ (ii):

Let $p = \pi\overline{\pi}$ for some Gauss prime $\pi = a + bi$ which implies $p = a^2 + b^2$.

(ii) $\Rightarrow$ (i):

If $p = a^2 + b^2$, then $p = (a + bi)(a - bi)$ provides a factorization of p in the ring of Gauss integers which is a prime factorization because of Theorem 9.1.

Thus it is established that (i) and (ii) are equivalent. Since it follows from Theorem 7.2 that (ii) and (iii) are equivalent, we are through. □

Remark 9.1. From Theorem 9.1 and the equivalence of (i) and (iii) in Theorem 9.3, it follows that the prime integers which are Gauss primes are those congruent to 3 modulo 4.

One observes that the Gaussian integers are the points of a square lattice in the complex plane. Similarly, the ring $\mathbf{Z}[\sqrt{-5}]$, consisting of complex numbers of the form $a + b\sqrt{-5}$, $a, b \in \mathbf{Z}$. is also an example of a Lattice in the complex plane. It is clear that non-zero ideals of these rings are sublattices.

9.2 The ring $\mathbf{Z}[\sqrt{-5}]$

We move on to discuss about the ring $\mathbf{Z}[\sqrt{-5}]$, which, as had been mentioned earlier, is not a UFD. Arguing as in the case of $\mathbf{Z}[i]$, with the function norm $=$ (absolute value)^2 : $\mathbf{Z}[\sqrt{-5}] \to \mathbf{Z}$, given by $a + b\sqrt{-5} \mapsto a^2 + 5b^2$, one observes that the units in $\mathbf{Z}[\sqrt{-5}]$ are those having norm 1 and hence they are ± 1. Since norm is multiplicative, a proper divisor of $1 + \sqrt{-5}$ or $1 + \sqrt{-5}$ must have its norm equal to a proper divisor of 6, that is 2 or 3. Since the ring $\mathbf{Z}[\sqrt{-5}]$ does not have any such element, it follows that the elements $1 + \sqrt{-5}$ and $1 + \sqrt{-5}$ are irreducible elements of $\mathbf{Z}[\sqrt{-5}]$.

We are now in a position to look back at Remark 4.1 and understand that $6 = 2 \times 3 = (1 + \sqrt{-5})(1 - \sqrt{-5})$ does indeed give two essentially different factorizations of an element in $\mathbf{Z}[\sqrt{-5}]$.

Thus, the ring $\mathbf{Z}[\sqrt{-5}]$ is not a UFD. Theorem 4.3 then implies that $\mathbf{Z}[\sqrt{-5}]$ is not a PID. The remaining part of this chapter is devoted to a characterization of the non-principal ideals of $\mathbf{Z}[\sqrt{-5}]$. In doing so, we closely follow the treatment in [Ar 1994].

Theorem 9.4. Let r be the minimum attained by the absolute values of the non-zero elements of a non-zero ideal $\mathtt{A}$ of the ring $\mathbf{Z}[\sqrt{-5}]$. Let $\gamma \in \mathtt{A}$ and D be the disc on the complex plane with centre $\frac{1}{n}\gamma$ and radius $\frac{1}{n}r$ for some positive integer n. Then the interior of D does not contain any point of $\mathtt{A}$ except possibly the centre $\frac{1}{n}\gamma$.

Proof: Let β be a point in the interior of D. This means that $|n\beta - \gamma| < r$. Now, if $\beta \in \mathtt{A}$, then $n\beta - \gamma$ is an element of $\mathtt{A}$ of absolute value less than r. This forces $n\beta - \gamma$ to be zero which proves the theorem. □

Now, let $\mathtt{A}$ be a non-zero ideal of the ring $\mathbf{Z}[\sqrt{-5}]$ and α a non-zero element of $\mathtt{A}$ of minimal absolute value r. The principal ideal (α) consists of complex numbers $(a+b\sqrt{-5})\alpha$ where $a, b \in \mathbf{Z}$. So the principal ideal has lattice basis $(\alpha, \alpha\sqrt{-5})$. If $\mathtt{A} > (\alpha)$, let β be an element of $\mathtt{A}$ which is not in (α). The element β may be chosen so as to lie in the rectangle with vertices $0, \alpha, \alpha\sqrt{-5}, \alpha + \alpha\sqrt{-5}$. We consider four discs each of radius r about the four vertices of the rectangle. We consider three more discs of radius $r/2$ each about the points $(\alpha\sqrt{-5})/2, (\alpha + \alpha\sqrt{-5})/2$ and $\alpha + (\alpha\sqrt{-5})/2$. We observe that these seven discs cover the rectangle. According to Theorem 9.4, the only points of the interiors of the discs which can lie in $\mathtt{A}$ are the centres of the discs. Hence, β has to be one of the half lattice points $(\alpha\sqrt{-5})/2, (\alpha + \alpha\sqrt{-5})/2$ and $\alpha + (\alpha\sqrt{-5})/2$. If $(\alpha\sqrt{-5})/2 \in \mathtt{A}$, then multiplying by $\sqrt{-5}$, $5\alpha/2 \in \mathtt{A}$. Since $\alpha \in \mathtt{A}$, this would imply $\alpha/2 \in \mathtt{A}$ which contradicts the choice of α. Since $(\alpha\sqrt{-5})/2 \notin \mathtt{A}$, $\alpha + (\alpha\sqrt{-5})/2$ too can not be in $\mathtt{A}$. Hence, $(\alpha + \alpha\sqrt{-5})/2 \in \mathtt{A}$.

To sum up,

Theorem 9.5. If $\mathtt{A}$ is a non-zero ideal of the ring $\mathbf{Z}[\sqrt{-5}]$ and α a non-zero element of $\mathtt{A}$ of minimal absolute value r, then either $\mathtt{A}$ is the principal ideal (α) with lattice basis $(\alpha, \alpha\sqrt{-5})$ or $\mathtt{A}$ is not a principal ideal and has the lattice basis $(\alpha, (\alpha + \alpha\sqrt{-5})/2)$. The second case occurs only when $(\alpha + \alpha\sqrt{-5})/2$ is an element of $\mathtt{A}$.

Remark 9.2. By Theorem 9.5 above, the ideal $(2, 1+\sqrt{-5})$ is an example of an ideal in $\mathbf{Z}[\sqrt{-5}]$ which is not principal.

Chapter 10

Algebraic number fields-I

The ring of Gaussian integers and the ring $\mathbf{Z}[\sqrt{-5}]$ we discussed in the last chapter, will be seen to belong to a special class of rings. More precisely, they are examples of rings of integers in algebraic number fields which will be our object of discussion now. The examples seen in the last chapter showed that unique factorization may or may not be available in such rings. However, in the next chapter, we shall see that there will be 'unique factorization in the ideal level' in all such rings.

10.1 Integral dependence

We start with some definitions and results bearing to a somewhat general situation.

Definition. Let B be a ring and A a subring of B. An element $\alpha \in B$ is said to be *integral* over A if α is a root of a monic polynomial in $A[x]$. If all the elements of B are integral over A, then B is said to be *integral* over A. A complex number α is said to be an *algebraic integer* if α is integral over $\mathbf{Z}$.

Exercise 10.1. Show that the set of algebraic integers in $\mathbf{Q}$ is $\mathbf{Z}$. In general, if A is a UFD, then show that the elements in the quotient field of A which are integral over A are precisely the elements of A.

Exercise 10.2. Let $n \in \mathbf{Z} \setminus \{0, 1\}$ be a square-free number with $n \equiv 1 \pmod 4$. Show that the ring $\mathbf{Z} + \mathbf{Z}(\sqrt{n})$ is not a UFD.

Theorem 10.1. Let B be a ring and A a subring of B. Then the following are equivalent:

(I) $\alpha \in B$ is integral over A.

(II) $A[\alpha]$ is a finitely generated A-module.

(III) $A[\alpha]$ is contained in a subring C of B such that C is a finitely generated A module.

(IV) There exists a faithful $A[\alpha]$-module M which is finitely generated as an A-module.

Proof:

I $\Rightarrow$ II.

α satisfies an equation $\alpha^n + a_1\alpha^{n-1} + \cdots + a_n = 0$, where a_i's are elements of A. Clearly, the A-module $A[\alpha]$ is generated by $1, \alpha, \cdots, \alpha^{n-1}$.

II $\Rightarrow$ III.

We can take $A[\alpha]$ as our C.

III $\Rightarrow$ IV.

C is a faithful $A[\alpha]$-module. Because, $aC = 0$ would imply $a.1 = 0$. Therefore we can take $M = C$.

IV $\Rightarrow$ I.

We consider the finitely generated A-module M and the A-module homomorphism $\phi : M \to M$ given by $\phi(\beta) = \alpha\beta$ for all $\beta \in M$. Since M is an $A[\alpha]$ module, we have $\alpha M \subset M$.

Therefore, by Theorem 8.1, $\alpha^n + a_1\alpha^{n-1} + \cdots + a_n \in \mathrm{Ann}_A(M)$ for some a_i's in A. Since M is faithful, $\alpha^n + a_1\alpha^{n-1} + \cdots + a_n = 0$. □

Corollary 10.1.1. If $b_1, b_2, \cdots, b_r$ are elements of B each integral over A, then $A[b_1, b_2, \cdots, b_r]$ is a finitely generated A-module.

Proof: Let B be a ring and A a subring of B such that B is a finitely generated A-module with a set of generators $\beta_1, \beta_2, \cdots, \beta_t$. If M is a finitely generated B-module such that the set $\{m_1, m_2, \cdots m_s\}$ generates M over B, then it is easy to see that the products $\beta_i m_j$, $1 \leq i \leq t, 1 \leq j \leq s$ generate M as an A-module and hence M is a finitely generated A-module.

With this observation, the corollary follows by induction on r. □

Corollary 10.1.2. The set C of elements of B which are integral over A form a subring of B.

Proof: If c_1, c_2 are two elements of C then $A[c_1, c_2]$ is a finitely generated A-module by the above corollary. Hence by (III) of the above theorem, $c_1 \pm c_2$ and $c_1 c_2$ are integral over A. $\square$

Corollary 10.1.3. If $A \subset B \subset C$ are rings such that B is integral over A and C is integral over B, then C is integral over A.

Proof: Let $c \in C$. Then $c^n + b_1 c^{n-1} + \cdots + b_n = 0$ where b_i's are in B. Let $B_1 = A[b_1, b_2, \cdots, b_n]$. Then B_1 is a finitely generated A-module by Corollary 10.1.1. Also, c being integral over B_1, $B_1[c]$ is a finitely generated B_1-module. Hence $B_1[c]$ is a finitely generated A-module and therefore c is integral over A. $\square$

Definition. The set C of elements of B which are integral over A is called *the integral closure* of A in B. From Corollary 10.1.2 we know that C is a ring. If $C = A$, then A is said to be *integrally closed* in B. If an integral domain A is integrally closed in its quotient field F, then we simply say that A is *integrally closed.*

Remark 10.1. In the above terminology, Exercise 10.1 says that a UFD is integrally closed.

10.2 Integers in Number Fields

At this point, we leave the territory of integral extensions of general rings and continue the discussion related to the particular case of algebraic number fields. We start with some preliminaries on number fields.

Definition. By an *algebraic number field* we mean a subfield K of $\mathbf{C}$ such that K is a finite extension of $\mathbf{Q}$. The integer $[K : \mathbf{Q}]$ will be called the *degree* of K (over $\mathbf{Q}$).

Remark 10.2. From Exercise 6.10 and Theorem 6.3, we see that for an algebraic number field K there is some $\theta \in K$ such that $K = \mathbf{Q}(\theta)$.

Remark 10.3. Let K be an algebraic number field of degree n. By Remark 10.2 and Exercise 6.2, $K = \mathbf{Q}(\theta)$ for some $\theta \in K$ whose minimal polynomial f in $\mathbf{Q}[x]$ is of degree n. If $\theta_1 (= \theta), \theta_2, \cdots, \theta_n$ are all the distinct roots of f, then for each $\theta_i (i = 1, 2, \cdots, n)$, there is an isomorphism σ_i of $K = \mathbf{Q}(\theta)$ onto $\mathbf{Q}(\theta_i) \subset \mathbf{C}$ defined by $\sigma_i(\sum_{j=0}^{m} b_j(\theta^j)) = \sum_{j=0}^{m} b_j(\theta_i^j)$. Clearly, σ_i's are all distinct and these are the only isomorphisms of K into $\mathbf{C}$.

Definition. The fields $K^{(i)} = \sigma_i(K)$, $i = 1, 2, \cdots, n$ in the above remark are called *conjugates* of K. If $K^{(i)} \subset \mathbf{R}$, then we call it a *real conjugate* of K; otherwise it is called a *complex conjugate.* Complex conjugates will occur in pairs; r_1 and $2r_2$ will respectively denote the numbers of real and complex conjugates of K. Also, for $\alpha \in K$, we shall write $\alpha^{(i)}$ for $\sigma_i(\alpha)$.

Theorem 10.2. Let K be an algebraic number field and $\omega_1, \omega_2, \cdots, \omega_n$ be a base of K over $\mathbf{Q}$. With the above notations, if Ω denotes the $n \times n$ matrix $[\omega_j^{(i)}]_{i,j}$, then Ω is nonsingular.

Proof: Let $K = \mathbf{Q}(\theta)$ as in Remark 10.3. Let A be the matrix corresponding to Ω for the base $1, \theta, \theta^2, \cdots, \theta^{n-1}$. Then, $\det A$ is the Vandermonde determinant and is equal to $\pm \prod_{1 \leq i < j \leq n} (\theta^{(i)} - \theta^{(j)}) \neq 0$. If B is the matrix which expressess (ω_i) in terms of the base $1, \theta, \theta^2, \cdots, \theta^{n-1}$, then $\Omega = BA$. Now, B being a matrix which has an inverse with entries in $\mathbf{Q}$, the result follows. □

We now embark upon the study of the ring of algebraic integers in an algebraic number field. If K is an algebraic number field, then the ring of algebraic integers in K will be denoted by $\mathbf{O}_K$.

Remark 10.4. For any algebraic number α, there exists $m \neq 0$ in $\mathbf{Z}$ such that $m\alpha$ is an algebraic integer. Because, if $a_n\alpha^n + a_{n-1}\alpha^{n-1} + \cdots + a_0 = 0$, where a_i's are in $\mathbf{Z}$, then $(a_n\alpha)^n + a_{n-1}(a_n\alpha)^{n-1} + \cdots + a_0 \cdot a_n^{n-1} = 0$, which implies that $a_n\alpha$ is an algebraic integer.

Definition. Let K be an algebraic number field and $\omega_1, \omega_2, \cdots, \omega_n$ be a base of K over $\mathbf{Q}$. For any $\alpha \in K$, the mapping $x \mapsto \alpha x$ is a linear mapping of K into itself considered as a $\mathbf{Q}$-vector space. The *trace* of α, denoted by $Tr(\alpha)$,$Tr_K(\alpha)$, or $Tr_{K/\mathbf{Q}}(\alpha)$, is defined to be the trace of the above linear map. Similarly, the *norm* of α, denoted by $N(\alpha)$, $N_K(\alpha)$ or $N_{K/\mathbf{Q}}(\alpha)$, is the determinant of that linear map. Clearly, $Tr_K(\alpha)$ and $N_K(\alpha)$ are in $\mathbf{Q}$.

Remark 10.5. For $\alpha \in K$, let $\alpha\omega_j = \sum_{i=1}^{n} a_{ij}\omega_i$ for $j = 1, \cdots, n$. Let $A = [a_{ij}]$. With the notations introduced just before Theorem 10.2, $(\alpha\omega_j)^{(k)} = \alpha^{(k)}\omega_j^{(k)} = \sum_{i=1}^{n} a_{ij}\omega_i^{(k)}$ and therefore, denoting the diagonal matrix $(\alpha^{(i)}\delta_{ij})$ by A_0, we have $A_0\Omega = \Omega A$. By Theorem 10.2, Ω has an inverse Ω^{-1}

and therefore, $A_0 = \Omega A \Omega^{-1}$. Hence, $N_K(\alpha) = \det A = \det(\Omega A \Omega^{-1}) = \det A_0 = \alpha^{(1)} \cdots \alpha^{(n)}$. Similarly, $Tr_K(\alpha) = TrA = Tr(\Omega A \Omega^{-1}) = TrA_0 = \alpha^{(1)} + \cdots + \alpha^{(n)}$. This gives alternative definitions for $Tr_K(\alpha)$ and $N_K(\alpha)$.

Remark 10.6. If A and B are the matrices for the linear maps corresponding to the elements $\alpha, \beta \in K$ respectively, then $A+B$ and AB are the matrices corresponding to $\alpha + \beta$ and $\alpha\beta$ respectively. Hence the mapping $\alpha \mapsto A$ is a homomorphism of K into the space of $n \times n$ - matrices over $\mathbf{Q}$. It is called a *regular representation* of K corresponding to the given base $\omega_1, \omega_2, \cdots, \omega_n$. It also follows that the trace of the sum of two elements of K is the sum of the traces of the elements. Similarly, the norm of the product of two elements of K is the product of the norms of the elements.

Remark 10.7. Let K be an algebraic number field of degree n. Let α be an algebraic integer in K. If α is of degree $m(\leq n)$ over $\mathbf{Q}$, then $1, \alpha, \cdots \alpha^{m-1}$ will be a $\mathbf{Q}$- basis for $\mathbf{Q}(\alpha)$. Let A be the $m \times m$-matrix with entries in $\mathbf{Q}$ which correspond to α in the regular representation of $\mathbf{Q}(\alpha)$, with respect to the basis $1, \alpha, \cdots, \alpha^{m-1}$. Let $\beta_1, \beta_2, \cdots, \beta_l$ be a basis of K considered as a vector space over $\mathbf{Q}(\alpha)$. Then $n = l.m$ and $\beta_1, \beta_1\alpha, \cdots, \beta_1\alpha^{m-1}, \beta_2, \beta_2\alpha, \cdots, \beta_2\alpha^{m-1}, \cdots, \beta_l, \beta_l\alpha, \cdots, \beta_l\alpha^{m-1}$ will constitute a basis of K over $\mathbf{Q}$. Let A_1 be the matrix which correspond to α in the regular representation of K with respect to this $\mathbf{Q}$-basis. Then

$$A_1 = \begin{vmatrix} A & 0 & \ldots 0 \\ 0 & A & \ldots 0 \\ 0 & 0 & \ldots A \end{vmatrix}$$

It follows that $Tr(A_1) = l.Tr(A)$. Since all the elements of A are in $\mathbf{Z}$, if $x^m + a_{m-1}x^{m-1} + \cdots + a_0$ is the minimal polynomial for α, we get that $Tr(A)$ and $Tr(A_1) = l.Tr(A) = -l.a_{m-1}$ are integers. Thus, for an algebraic integer $\alpha \in K$, $Tr_K(\alpha)$ is an integer. Similarly, $N_K(\alpha) = \det A_1 = (\det A)^l \in \mathbf{Z}$.

Theorem 10.3 The bilinear form $B(x,y) \stackrel{\text{def}}{=} Tr_K(xy)$ for $x, y \in K$ is non-degenerate.

Proof: Let $x \neq 0$ be in K. Then fixing x, $Tr_K(xy)$ is not identically zero in y, because for $y = x^{-1}$, $Tr_K(xy) = Tr_K(1) = n$. Similarly for $y \neq 0$ in K, $Tr_K(xy)$ is not identically zero in x. □

The following corollary follows from Exercise 8.6.

Corollary 10.3.1. To any **Q**-base $\omega_1, \omega_2, \cdots, \omega_n$ of K, there corresponds a base $\omega'_1, \omega'_2, \cdots, \omega'_n$ such that $Tr_K(\omega_i \omega'_j) = \delta_{ij},\ 1 \leq i, j \leq n$.

Theorem 10.4. Let K be an algebraic number field of degree n and $\mathbf{O}_K$, the ring of algebraic integers in K. Then there exists a **Q**-base $\omega_1, \omega_2, \cdots, \omega_n$ of K such that ω_i's are in $\mathbf{O}_K$ and $\mathbf{O}_K = \mathbf{Z}\omega_1 + \cdots + \mathbf{Z}\omega_n$.

Proof: Since for every element α of K, there is an $m(\neq 0) \in \mathbf{Z}$ such that $m\alpha \in \mathbf{O}_K$, there is a **Q**-base $v_1, \cdots, v_n$ of K consisting of elements of $\mathbf{O}_K$. Let $v'_1, \cdots, v'_n$ be a base of K such that

$$Tr_K(v_i v'_j) = \delta_{ij},\ 1 \leq i, j \leq n. \tag{10.1}$$

For any $z \in \mathbf{O}_K$, $z = \sum_{i=1}^{n} a_i v'_i$ with a_i's belonging to **Q**. Now, $zv_i \in \mathbf{O}_K$ for $1 \leq i \leq n$. Therefore, $Tr_K(zv_i) \in \mathbf{Z}$. But, by (10.1), $Tr_K(zv_i) = a_i$ and hence $\mathbf{O}_K \subset \mathbf{Z}v'_1 + \cdots + \mathbf{Z}v'_n$. Hence there exist $\omega_1, \omega_2, \cdots, \omega_m$ $(m \leq n)$ such that $\mathbf{O}_K = \mathbf{Z}\omega_1 + \cdots + \mathbf{Z}\omega_m$. But, if $m < n$, then the **Q**-subspace of K generated by $\omega_1, \omega_2, \cdots, \omega_m$ being K itself, it will contradict the fact that the dimension of K over **Q** is n. Hence $m = n$. Also, $\omega_1, \omega_2, \cdots, \omega_n$ have to be **Q**-independent and the result follows. □

Definition. The elements $\omega_1, \omega_2, \cdots, \omega_n$ in the above theorem is called an *integral basis* of K.

In Exercise A.9 we have seen that every non-zero ideal in the ring of Gaussian integers contains a non-zero integer. The following remark gives a general statement to that effect.

Remark 10.8. Let K be an algebraic number field. Let $\mathtt{A}$ be any non-zero ideal in $\mathbf{O}_K$. We claim that $\mathtt{A} \cap \mathbf{Z} \neq \{0\}$. For, if $\alpha \neq 0$ is an algebraic integer in $\mathtt{A}$, then $\alpha^r + a_{r-1}\alpha^{r-1} + \cdots + a_0 = 0$ for a_i's in $\mathbf{Z}$ such that $a_0 \neq 0$. Then clearly $a_0 = -\alpha(a_1 + \cdots + \alpha^{r-1}) \in \mathtt{A}$.

Also, from Remark 10.4, we know that for any $\alpha \in K$, there exists $m \neq 0$ in $\mathbf{Z}$ such that $m\alpha \in \mathbf{O}_K$. Since $\mathtt{A}$ has an element $m_1(\neq 0) \in \mathbf{Z}$ in it, $mm_1\alpha \in \mathtt{A}$. Thus, for any $\alpha \in K$, there exists $l \neq 0$ in $\mathbf{Z}$ such that $l\alpha \in \mathtt{A}$

Remark 10.9. Let $\mathtt{A}$ be a non-zero ideal in $\mathbf{O}_K$. If $\omega_1, \omega_2, \cdots, \omega_n$ are as in Theorem 10.4, then $\mathtt{A} \subset \mathbf{O}_K = \mathbf{Z}\omega_1 + \cdots + \mathbf{Z}\omega_n$. Therefore, since the ring $\mathbf{Z}$ is Noetherian, by Corollary 8.4.2 and Theorem 8.3, there exists $v_1, \cdots, v_m$ $(m \leq n)$ in $\mathtt{A}$ such that $\mathtt{A} = \mathbf{Z}v_1 + \cdots + \mathbf{Z}v_m$. From Remark 10.8, $v_1, \cdots, v_m$ generate K over $\mathbf{Q}$ and hence we must have $m = n$. The v_i's are said to constitute an *integral basis* of $\mathtt{A}$.

Further, we may choose the v_i's in such a way that $v_i = \sum_{j \geq i} p_{ij}\omega_j$, $p_{ij} \in \mathbf{Z}$.

Remark 10.10. If $\mathtt{A}$ is a non-zero ideal of $\mathbf{O}_K$, then by Remark 10.8, there exists $0 \neq a \in \mathbf{Z}$ such that $a\mathbf{O}_K \subset \mathtt{A} \subset \mathbf{O}_K$. Now if $\mathbf{O}_K = \mathbf{Z}\omega_1 + \cdots + \mathbf{Z}\omega_n$, then $a\mathbf{O}_K = \mathbf{Z}a\omega_1 + \cdots + \mathbf{Z}a\omega_n$ so that the number of elements in $\mathbf{O}_K/a\mathbf{O}_K$ is a^n. Therefore $\mathbf{O}_K/\mathtt{A}$ is also finite. The number of elements in the residue class ring $\mathbf{O}_K/\mathtt{A}$ is called the *norm* of $\mathtt{A}$ and is denoted by $N(\mathtt{A})$. If $\mathtt{A} = \{0\}$, we define $N(\mathtt{A}) = 0$.

Remark 10.11. If $\wp$ is a prime ideal in $\mathbf{O}_K$, then from Remark 10.8, we conclude that $\wp$ contains one prime number $p > 0$ of $\mathbf{Z}$. Now, if p and q are distinct primes in $\wp$, then we can find integers x, y in $\mathbf{Z}$ such that $xp + yq = 1$ which would imply that $\wp = \mathbf{O}_K$, which is a contradiction. Hence, $\wp$ contains exactly one prime number $p > 0$ of $\mathbf{Z}$.

To distinguish from the prime ideals in $\mathbf{O}_K$, a prime number $p \in \mathbf{Z}$ will be referred to as a *rational prime*.

Theorem 10.5. The ring of algebraic integers $\mathbf{O}_K$ in an algebraic number field K has the following properties:

i) every non-zero prime ideal of $\mathbf{O}_K$ is maximal.

ii) $\mathbf{O}_K$ is integrally closed.

iii) $\mathbf{O}_K$ is Noetherian.

Proof: Let K be an algebraic number field of degree n and $\mathbf{O}_K$, the ring of algebraic integers in K. Let $\wp$ be a non-zero prime ideal in $\mathbf{O}_K$. Then $\mathbf{O}_K/\wp$ is a finite integral domain and hence a field. This proves that every non-zero prime ideal of $\mathbf{O}_K$ is maximal. Again, from Corollary 10.1.3, it follows that the integral closure of $\mathbf{O}_K$ in K is $\mathbf{O}_K$ itself. Finally, we observe that for ideals $\mathtt{A}, \mathtt{B}$ of $\mathbf{O}_K$ such that $\mathtt{A} \subset \mathtt{B}$ and $\mathtt{A} \neq \mathtt{B}$, we have $N(\mathtt{A}) > N(\mathtt{B})$. Hence if $\mathtt{A}_1 \subset \mathtt{A}_2 \subset \cdots \subset \mathtt{A}_n \subset \cdots$ is an increasing sequence of ideals in $\mathbf{O}_K$, then $\mathtt{A_m} = \mathtt{A_{m+1}}$ for $m > m_0$ $(> 0) \in \mathbf{Z}$. That is, $\mathbf{O}_K$ is Noetherian. □

Exercise set C

C.1 Let a, b, c be ordinary integers such that they do not have any common factor and $a^2 + b^2 = c^2$. If π is a Gaussian prime which divides $a + ib$ in the ring of Gaussian integers, then show that the power of π which appears in the unique factorization of $a + ib$ into prime powers, is even.

C.2 Find all integer solutions x, y, z of $x^2 + y^2 = z^2$ having no common factor.

C.3 Determine all prime integers p such that $x^2 + 5y^2 = p$ has an integer solution.

C.4 Let ω denote the number $(-1 + \sqrt{-3})/2$ and D the domain $\mathbf{Z}[w]$.

Then, show that

(a) D is a Euclidean domain with the size function $N(a + b\omega) = a^2 - ab + b^2$.

(b) $\alpha \in D$ is a unit iff $N(\alpha) = 1$. The units are $\pm 1, \pm\omega, \pm\omega^2$.

(c) If π is a prime in D, then there is a rational prime p such that $N(\pi) = p$ or p^2. In the first case, π is a prime in D.

(d) If q is a rational prime such that $q \equiv 2 \pmod 3$, then q is a prime in D. If q is a rational prime such that $q \equiv 1 \pmod 3$, then $q = \pi\overline{\pi}$ where π is a prime in D.

(e) $(1 - \omega)$ is a prime in D and $3 = -\omega^2(1 - \omega)^2$.

Chapter 11

Dedekind domains

In this chapter, we shall learn some of the fundamental properties of Dedekind domains. The collection of Dedekind domains will contain the principal ideal domains for instance. The rings of integers in algebraic number fields, the objects of our main interest, will supply non-trivial examples of Dedekind domains.

11.1 Fractional ideals

Let R be an integral domain and K its quotient field.

Definition. By a *fractional ideal* of R, we shall mean a non-zero R-module A contained in K such that $mA \subset R$ for some $m \neq 0$ in R.

Remark 11.1. An ideal in R is trivially a fractional ideal; we shall call it an *integral ideal.* Also, any fractional ideal A is clearly of the form $a^{-1}B$ for some $a \neq 0$ in R and an integral ideal B.

Definition. For a fractional ideal A of R, by A' we shall denote the set

$$\{x \in K : xA \subset R\}.$$

Exercise 11.1. Show that for a fractional ideal A of R, A' defined above is also a fractional ideal.

If M is an R-module and A an ideal of R, then we have already defined the product AM. We define the product of two fractional ideals A and B of

R in the same way. That is, AB is the set of all finite sums $a_1b_1 + \cdots + a_rb_r$ where $a_i \in A$ and $b_i \in B$. It is easy to check that AB is also a fractional ideal of R.

Definition. For a fractional ideal A of R, clearly $AA' \subset R$. If for a fractional ideal A, we have $AA' = R$, then A is called *invertible* and we write $A' = A^{-1}$.

For $a(\neq 0) \in K$, aR is clearly a fractional ideal. Such fractional ideals are called *principal fractional ideals.*

Remark 11.2. One observes that a principal fractional ideal is invertible.

Exercise 11.2. Show that the set of all invertible fractional ideals of an integral domain R form a group under multiplication.

Definition. If R is an integral domain such that every fractional ideal of R is invertible, then R is called a *Dedekind domain.*

Remark 11.3. From Remarks 11.1 and 11.2, we see that a principal ideal domain is a Dedekind domain.

11.2 Properties of Dedekind domains

In the next three theorems we establish some important properties of Dedekind domains.

Theorem 11.1. A Dedekind domain is Noetherian.

Proof: Let A be any non-zero ideal of a Dedekind domain R. Since A is invertible, there are elements $a_i \in A$ and $b_i \in A^{-1}$, for $i = 1, 2, \cdots, n$ such that $a_1b_1 + \cdots + a_nb_n = 1$. Now, given any element α of A, we have $\alpha = (\alpha b_1)a_1 + \cdots + (\alpha b_n)a_n$ where $\alpha b_i \in R$ for all $i = 1, \cdots, n$. Thus, a_i's generate A. Hence R is Noetherian. □

Theorem 11.2. In a Dedekind domain every non-zero prime ideal is maximal.

Proof: Let P be any non-zero prime ideal of a Dedekind domain R. Let M be a maximal ideal containing P.

Now, $PM^{-1} \subset MM^{-1} = R$. This shows that PM^{-1} is an ideal of R. Since $(PM^{-1})M = P$, where P is a prime ideal, either $PM^{-1} \subset P$ or $M \subset P$. If $PM^{-1} \subset P$, then $M^{-1} \subset P^{-1}P = R$. Since $R \subset M^{-1}$ by definition, we have $M^{-1} = R$, which would imply $M = R$ – contradiction. Hence $M = P$ and this proves the theorem. □

Theorem 11.3. A Dedekind domain is integrally closed.

Proof: Let R be a Dedekind domain and K its quotient field. Let α be an element of K such that α is integral over R. Then, by Theorem 10.1, $R[\alpha]$ is a finitely generated R-module. Let $a_1, a_2 \cdots, a_n$ generate the R-module $R[\alpha]$.

K being the quotient field of R, there is $b\ (\neq 0) \in R$ such that $ba_i \in R$ for $i = 1, 2, \cdots, n$. This implies that $bR[\alpha] \subset R$. Hence $R[\alpha]$ is a fractional ideal.

$$\text{Now,}\quad R[\alpha] = RR[\alpha] = R[\alpha]^{-1}R[\alpha]R[\alpha] = R[\alpha]^{-1}R[\alpha] = R.$$

Hence $\alpha \in R$ and we are through. □

The next few theorems will lead us to the fact (see Remark 11.4) that the properties of a Dedekind domain which we have obtained in Theorems 11.1–11.3, do indeed characterize a Dedekind domain.

Theorem 11.4. Let R be a Noetherian domain. Then given any non-zero proper ideal A of R, one can find prime ideals $P_1, P_2 \cdots, P_m$ of R such that

$$P_1P_2\cdots P_m \subset A \subset P_1 \cap P_2 \cap \cdots P_m.$$

Proof: If possible, let there be non-zero proper ideals of R which do not satisfy the property. Let I be a maximal element of the set of such ideals. Clearly, this ideal can not be prime. Therefore, there are elements a, b of R such that $a, b \notin I$ and $ab \in I$. Let $A = I + aR$ and $B = I + bR$. Since $AB \subset I$, $A = R$ would imply $B = I$ which can not be true in view of the fact that $b \notin I$. Therefore, $A \neq R$. Similarly, $B \neq R$. Therefore, A and B satisfy the property mentioned in the theorem. Since by construction, $AB \subset I \subset A \cap B$, I also has the property, which is a contradiction. □

Theorem 11.5. Let an integral domain R satisfy the following conditions:

(i) R is Noetherian.

(ii) R is integrally closed.

(iii) every non-zero prime ideal of R is maximal.

Then every non-zero prime ideal of R is invertible.

Proof: Let P be a non-zero prime ideal of R. Let $a(\neq 0) \in P$. By Theorem 11.4 above, we can find the least positive integer m such that the principal ideal aR contains a product $P_1P_2 \cdots P_m$ of m prime ideals. Since P is a prime ideal, by condition (iii) we see that P has to be equal to one of the prime ideals $P_1, P_2, \cdots, P_m$. Without loss of generality, let $P = P_1$. By the minimality of m, the product $P_2 \cdots P_m$ is not contained in aR. Let $b \in P_2 \cdots P_m \setminus aR$. Thus $ba^{-1} \notin R$. Now, $bP \subset PP_2 \cdots P_m \subset aR$, which would imply $ba^{-1}P \subset R$. Thus $ba^{-1} \in P' \setminus R$. This proves that $R < P'$.

Let x be any non-zero element of P. By the observation in the previous paragraph, there exists $y \in P' \setminus R$.

Now, $P = RP \subset P'P \subset R$. By (iii), either $P'P = P$ or $P'P = R$.

If $P'P = P$, then $(P')^nP = P$ for any positive integer n. Consequently, $xy^n \in P$ for all integers $n \geq 1$. Thus $xR[y] \subset R$. Hence $xR[y]$ is an ideal of R and by (i) it is finitely generated. Let $a_1, \cdots, a_m$ generate $xR[y]$ as an R- module. Then $a_1x^{-1}, \cdots, a_mx^{-1}$ generate $R[y]$ as an R- module. By Theorem 10.1, y is integral over R and hence by (ii), $y \in R$, contradicting the choice of y.

Therefore, $P'P = R$ and we are through. □

Theorem 11.6. Let R be an integral domain satisfying the conditions (i), (ii) and (iii) of the previous theorem. Then, any proper ideal R (i.e. ideals other than (0) or R) can be written as the product of prime ideals in R determined uniquely upto order.

Proof: First we prove the existence of a factorization. Let $\mathcal{S}$ be the set of all proper ideals of R which can not be factorized into prime ideals. If possible, let $\mathcal{S} \neq \emptyset$. Now, every element of $\mathcal{S}$ contains a product of prime ideals. We can choose an element A of $\mathcal{S}$ such that A contains a product $P_1P_2 \cdots P_m$ of prime ideals where m is minimal. By condition (iii), $m \neq 1$ as A can not be prime. There is a prime ideal $\wp$ ($\wp \neq A$) such that $\wp \supset A$. Then $\wp$ has to be equal to one of the prime ideals P_i's, say P_1. By the previous theorem, there is a fractional ideal $\wp^{-1}$ such that $\wp\wp^{-1} = R$. Now, $P_2 \cdots P_m \subset A\wp^{-1} \subset \wp\wp^{-1} = R$. By the minimality of m, $A\wp^{-1} = \wp_1\wp_2\cdots\wp_r$ for prime ideals $\wp_1, \wp_2, \cdots, \wp_r$. But, then $A = \wp\wp_1\wp_2\cdots\wp_r$, contradicting the choice of A.

Now, we proceed to prove the uniqueness of the factorization. If possible let a proper ideal $\mathtt{A} \subset R$ have two factorizations

$$\mathtt{A} = \mathtt{P}_1\mathtt{P}_2\cdots\mathtt{P}_r = \wp_1\wp_2\cdots\wp_s \tag{11.1}$$

where $\mathtt{P}_1, \mathtt{P}_2, \cdots, \mathtt{P}_r, \wp_1, \cdots, \wp_s$ are prime ideals. Since $\wp_1$ is prime, it contains one of the ideals $\mathtt{P}_1, \mathtt{P}_2, \cdots, \mathtt{P}_r$, say $\mathtt{P}_1$. Since, $\mathtt{P}_1$ is maximal, $\mathtt{P}_1 = \wp_1$. Multiplying (11.1) by ${\mathtt{P}_1}^{-1} = {\wp_1}^{-1}$ we have, $\mathtt{P}_2\cdots\mathtt{P}_r = \wp_2\cdots\wp_s$. We repeat the argument and in finitely many steps it follows that $r = s$ and the factorization is unique upto order. □

Corollary 11.6.1. Any non-zero fractional ideal $\mathtt{A}$ can be written uniquely in the form $\mathtt{P}_1\mathtt{P}_2\cdots\mathtt{P}_r{\wp_1}^{-1}{\wp_2}^{-1}\cdots{\wp_s}^{-1}$ where the prime ideals $\mathtt{P}_i$'s are distinct from the prime ideals $\wp_j$.

Proof: There is $c\ (\neq 0) \in R$ such that $c\mathtt{A}(= \mathtt{B}$ say$) \subset R$. Since cR and $\mathtt{B}$ can be written as the product of prime ideals in R determined uniquely upto order, the result follows. □

Remark 11.4. From Theorems 11.1, 11.2, 11.3, 11.5 and Corollary 11.6.1, we get that an integral domain R is a Dedekind domain if and only if it satisfies the three conditions in Theorem 11.5. Hence, Theorem 10.5 says that the ring of integers of an algebraic number field is a Dedekind domain.

Remark 11.5. The *greatest common divisor* $(\mathtt{A}, \mathtt{B})$ of two ideals $\mathtt{A}, \mathtt{B}$ of a Dedekind domain R is defined to be ${\wp_1}^{c_1}{\wp_2}^{c_2}\cdots{\wp_s}^{c_s}$ where $\wp_i$'s are the prime ideals which appear in the factorization of $\mathtt{A}$ or $\mathtt{B}$ and ${\wp_i}^{c_i}$ is the minimum of the powers of $\wp_i$ $(c_i \geq 0)$ which appear in the factorizations of the two ideals $\mathtt{A}$ and $\mathtt{B}$. If $\mathtt{A}, \mathtt{B}$ are two ideals of R, then $\mathtt{A} \subset \mathtt{B}$ if and only if $\mathtt{A} = \mathtt{BC}$ for some ideal $\mathtt{C}$ of R. For, if $\mathtt{A} \subset \mathtt{B}$, then $\mathtt{C} = \mathtt{AB}^{-1} \subset R$ and conversely, if $\mathtt{A} = \mathtt{BC}$ for some ideal $\mathtt{C}$, then $\mathtt{A} \subset \mathtt{B}$. Therefore, the greatest common divisor of $\mathtt{A}$ and $\mathtt{B}$ is clearly the smallest ideal containing both $\mathtt{A}$ and $\mathtt{B}$ which is nothing but $\mathtt{A} + \mathtt{B}$.

If for two ideals $\mathtt{A}, \mathtt{B}$ of a Dedekind domain R, their greatest common divisor $(\mathtt{A}, \mathtt{B}) = R$, that is, there is no prime ideal which divides both $\mathtt{A}$ and $\mathtt{B}$, then $\mathtt{A}$ and $\mathtt{B}$ are said to be relatively prime.

If I is an ideal of a ring R, then one writes $a \equiv b \pmod{I}$ to mean that $a - b \in I$.

Exercise 11.3. Let R be a Dedekind domain. Let P be any prime ideal of R. Then show that for any $b \in R$, $a \in R \setminus P$ and any positive integer n, the congruence $ax \equiv b \pmod{P^n}$ has a solution in x.

The next exercise states the natural extension of the *Chinese remainder theorem* to Dedekind domains.

Exercise 11.4. If $I_1, I_2, \cdots, I_m$ are pairwise relatively prime ideals of a Dedekind domain R and elements $a_1, a_2, \cdots, a_m$ in R are given, then there exists a common solution of the congruences

$$x \equiv a_i \pmod{I_i}, \qquad i = 1, 2, \cdots, m.$$

Exercise 11.5. Using the Chinese remainder theorem as above, show that a Dedekind domain with only finitely many prime ideals is a PID.

Exercise 11.6. Show that a Dedekind domain is a UFD if and only if it is a PID.

Remark 11.6. The ring $\mathbf{Z}[\sqrt{-5}]$, which is the ring of integers of the number field $\mathbf{Q}(\sqrt{-5})$ (see Chapter 13), is seen not to be a PID. Therefore, it will have infinitely many prime ideals. Since, each of them will contain exactly one rational prime and each rational prime p can belong to only finitely many prime ideals (namely those occuring in the unique factorization of the principal ideal (p)), we get yet another proof of the fact that there are infinitely many rational primes!

Chapter 12

Algebraic number fields-II

In this chapter, we shall mainly discuss about class number of a number field. We shall see that class number of a number field is a positive integer, which, in some sense, measures the deviation from unique factorization of the corresponding ring of integers. Towards the end of this chapter, we prove Dirichlet's unit theorem which gives the structure of the group of units of the ring of algebraic integers in an algebraic number field.

12.1 Class groups

Definition. From Corollary 11.6.1 and Remark 11.4, it follows that the non-zero fractional ideals in an algebraic number field K form a multiplicative group. We denote this group by $\Delta(K)$.

As had been already mentioned, a fractional ideal in K is said to be *principal* if it is of the form $\alpha\mathbf{O}_K$ for some $\alpha \in K$. Clearly the principal ideals form a subgroup $\Pi(K)$ of $\Delta(K)$. The quotient group $\Delta(K)/\Pi(K)$ is called the *group of ideal classes* or simply the *class group* of K. Two fractional ideals $\mathtt{A}$ and $\mathtt{B}$ $(\neq \{0\})$ in K are in the same ideal class or are equivalent if and only if $\mathtt{A} = (\alpha\mathbf{O}_K)\mathtt{B}$ for some $\alpha \in K$.

Theorem 12.1. Let $\mathtt{A} \neq \{0\}$ be an ideal of $\mathbf{O}_K = \mathbf{Z}\omega_1 + \cdots + \mathbf{Z}\omega_n$. Then there exist $\alpha_i = \sum_{j=i}^{n} p_{ij}\omega_j, p_{ii} > 0, p_{ij} \in \mathbf{Z}$ such that $\mathtt{A} = \mathbf{Z}\alpha_1 + \cdots + \mathbf{Z}\alpha_n$. Further $N(\mathtt{A}) = p_{11}p_{22}\cdots p_{nn}$.

Proof: The first part follows from Remark 10.9. That is, $\mathtt{A}$ has an integral basis $\alpha_1, \cdots, \alpha_n$ of the required form.

We claim that the $p_{11}p_{22}\cdots p_{nn}$ numbers

$$\eta(a_1, \cdots, a_n) = \sum_{i=1}^{n} a_i\omega_i, \quad 0 \leq a_i < p_{ii}, \; a_i \in \mathbf{Z}$$

form a complete system of residues of $\mathbf{O}_K$ modulo $\mathtt{A}$.

Let $\xi = \sum_{i=1}^{n} c_i\omega_i, c_i \in \mathbf{Z}$ be any element of $\mathbf{O}_K$. Let $\mathbf{O}_K^i = \mathbf{O}_K \cap (\mathbf{Z}\omega_{i+1} + \cdots + \mathbf{Z}\omega_n)$. Writing $c_1 = m_1p_{11} + a_1$ where $0 \leq a_1 < p_{11}$, we have $\xi - a_1\omega_1 - m_1\alpha_1 (= \xi_1, \text{say}) \in \mathbf{O}_K^1$. Proceeding in this way it is clear that $\xi = \sum_{i=1}^{n} m_i\alpha_i + \sum_{i=1}^{n} a_i\omega_i$, where $m_i, \; a_i \in \mathbf{Z}, \; 0 \leq a_i < p_{ii}$. Thus η's generate $\mathbf{O}_K$ modulo $\mathtt{A}$.

Next, if possible, suppose $\sum_{i=1}^{n} a_i\omega_i = \sum_{i=1}^{n} b_i\omega_i \pmod{\mathtt{A}}, \;\; 0 \leq a_i, b_i < p_{ii}$. Writing $a_i - b_i = d_i$, we get $\sum_{i=1}^{n} d_i\omega_i = \sum_{i=1}^{n} m_i\alpha_i, \;\; 0 \leq \mid d_i \mid < p_{ii}, \; m_i \in \mathbf{Z}$. But $\sum_{i=1}^{n} m_i\alpha_i = \sum_{i=1}^{n} c_i\omega_i$, where $c_1 = m_1p_{11}$. Since ω_i's are linearly independent, this implies that $d_1 = m_1p_{11}$ and since $0 \leq \mid d_i \mid < p_{ii}$, it therefore follows that $m_1 = d_1 = 0$. Now, $c_2 = m_2p_{22}$ and by above the argument, $d_2 = m_2 = 0$. Proceeding in this way it follows that $d_i = 0$ for all i. This proves that the η's are different modulo $\mathtt{A}$. □

Remark 12.1. Let $\mathtt{A}$ be an integral ideal in K and let $\mathtt{A} = \mathbf{Z}\alpha_1 + \cdots + \mathbf{Z}\alpha_n = \mathbf{Z}\beta_1 + \cdots + \mathbf{Z}\beta_n$ where α_i's and β_i's are in $\mathtt{A}$. Then $\alpha_i = \sum_{j=1}^{n} q_{ij}\beta_j$ and $\beta_i = \sum_{j=1}^{n} r_{ij}\alpha_j$ for $i = 1, \cdots, n$ where r_{ij}'s and q_{ij}'s are in $\mathbf{Z}$. Since the matrices $[r_{ij}]$ and $[q_{ij}]$ are inverses of each other, $\det[r_{ij}] = \det[q_{ij}] = \pm 1$.

Theorem 12.2. For $\alpha \neq 0$ in $\mathbf{O}_K$, then $N((\alpha)) = \mid N_K(\alpha) \mid$.

Proof: If $\mathbf{O}_K = \mathbf{Z}\omega_1 + \cdots + \mathbf{Z}\omega_n$, then taking $\mathtt{A} = \alpha\mathbf{O}_K$, by Theorem 12.1, there exist $\alpha_i = \sum_{j=1}^{n} p_{ij}\omega_j, p_{ii} > 0, p_{ii} \in \mathbf{Z}$ and $p_{ij} = 0$ for $1 \leq j < i \leq n$ such that $\mathtt{A} = \mathbf{Z}\alpha_1 + \cdots + \mathbf{Z}\alpha_n$ and $N(\mathtt{A}) = p_{11}p_{22}\cdots p_{nn}$. Writing P for the $n \times n$- matrix $[p_{ij}]$ and observing that $\alpha\mathbf{O}_K = \mathbf{Z}\alpha\omega_1 + \cdots + \mathbf{Z}\alpha\omega_n$ as well, if R is the matrix $[r_{ij}]$ where $\alpha\omega_i = \sum_{j=1}^{n} r_{ij}\alpha_j$ by Remark 12.1 we have $N_K(\alpha) = \det P. \det R = \pm \det P = \pm p_{11}p_{22}\cdots p_{nn}$. Hence the theorem. □

Corollary 12.2.1. If $t \in \mathbf{Z}$, then $N(t\mathbf{O}_K) = \mid N_K(t) \mid = \mid t^n \mid$.

Exercise 12.1. For two integral ideals $\mathtt{A}$ and $\mathtt{B}$ in an algebraic number field K, $N(\mathtt{AB}) = N(\mathtt{A})N(\mathtt{B})$.

Exercise 12.2. For any integer $m > 0$, the number of ideals $\mathtt{A} \subset \mathbf{O}_K$ for which $N(\mathtt{A}) \leq m$ is finite.

Theorem 12.3. There exists a constant C depending only on K such that every integral ideal $\mathtt{A} \neq 0$ will contain an $\alpha \neq 0$ with $| N_K(\alpha) | \leq CN(\mathtt{A})$.

Proof: Let $\mathbf{O}_K = \mathbf{Z}\omega_1 + \cdots + \mathbf{Z}\omega_n$ and t be the largest integer such that $t \leq (N(\mathtt{A}))^{1/n}$. Then among the $(t+1)^n$ numbers $\sum_{i=1}^n c_i\omega_i$ with $0 \leq c_i \leq t$, there will be at least two distinct numbers whose difference is in $\mathtt{A}$. Let $\alpha = a_1\omega_1 + \cdots + a_n\omega_n \neq 0$ be that difference. Then $| a_i | \leq t$ for all i. If A_i is the $n \times n$ matrix which corresponds to ω_i under the regular representation, then $A = \sum_{i=1}^n a_iA_i$ will be the matrix corresponding to α. Clearly elements of A are rational integers which are at most $t\mu$ in absolute value where the constant μ depends only on A_i's which in turn depend on the ω_i's i.e. μ depends only on K. Therefore $| N_K(\alpha) | = | \det A | \leq C \cdot t^n \leq C \cdot N(\mathtt{A})$ where the constant C depends only on K. □

Remark 12.2. In Theorem 12.10, the inequality in the statement of 12.3 will be replaced by one with an explicit constant in place of C.

Theorem 12.4. The class group of K is finite.

Proof: In view of Exercise 12.2, it will be enough to prove that in every class of ideals there is an integral ideal of norm $\leq C$ where C is a constant depending only on K. Let $\mathcal{S}$ be a class. Let $\mathtt{A}$ be an integral ideal in the inverse class $\mathcal{S}^{-1}$. By Theorem 12.3, there is $\alpha \in \mathtt{A}$ such that $| N_K(\alpha) | \leq C \cdot N(\mathtt{A})$ for a constant C depending only on K. Now, $\mathtt{B} = (\alpha)\mathtt{A}^{-1} \in \mathcal{S}$ and $N(\mathtt{A})N(\mathtt{B}) = N(\mathtt{AB}) = | N_K(\alpha) | \leq C \cdot N(\mathtt{A})$ which implies that $N(\mathtt{B}) \leq C$. This completes the proof. □

12.2 Discriminants

Let K be an algebraic number field of degree n. Let $\sigma_1, \cdots, \sigma_n$ be the distinct isomorphisms of K into $\mathbf{C}$. As before, r_1 and $2r_2$ are respectively the numbers of real and complex conjugates of K. Usually, we fix the order of $\sigma_1, \cdots, \sigma_n$ in such a way that $\sigma_1, \cdots \sigma_{r_1}$ are real and for the remaining embeddings, $\sigma_i = \overline{\sigma}_{i-r_2}$ for $i = r_1 + r_2 + 1, r_1 + r_2 + 2, \cdots n$.

Definition. The pair $[r_1, r_2]$ is called the *signature* of K.

We are now going to define and discuss about the absolute discriminant (denoted by $d(K)$) of an algebraic number field K. The absolute discriminant is one of the various functions introduced to obtain some kind of classification of algebraic number fields. The function $d(K)$ has the remarkable property of assuming the same value only for a finite number of fields.

First, we proceed to give some general definition.

Definition. Let K be a field of characteristic zero and L/K a finite extension of degree n. If $\overline{K}$ is a fixed algebraic closure of K , then there are n distinct embeddings $f_1, \cdots, f_n$ of L into $\overline{K}$ leaving K invariant.

For any $\nu_1, \cdots, \nu_n \in L$, we define the *discriminant* $d_{L/K}(\nu_1, \cdots, \nu_n)$ by $d_{L/K}(\nu_1, \cdots, \nu_n) = (\det[f_j(\nu_i)])^2$. We note that because of the squaring, the discriminant does not depend on the ordering of the embeddings or of the ν_i's.

Definition. If a is an arbitrary element of L, then its *discriminant* with respect to K (denoted by $d_{L/K}(a)$) is defined by

$$\begin{aligned} d_{L/K}(a) &= d_{L/K}(1, a, a^2, \cdots, a^{n-1}) \\ &= \Pi_{i<j}(f_i(a) - f_j(a))^2. \end{aligned}$$

Therefore, if a generates L over K, then $d_{L/K}(a)$ is nothing but the discriminant of the minimal polyniminal for a over K and is non-zero. If, however, a generates a proper subfield of L, then $d_{L/K}(a) = 0$.

Theorem 12.5. If $Tr_{L/K}(x) = f_1(x) + \cdots + f_n(x)$ is the trace defined for $x \in L$, then for every sequence $\nu_1, \cdots, \nu_n \in L$, we have

$$d_{L/K}(\nu_1, \cdots, \nu_n) = \det[Tr_{L/K}(\nu_i \nu_j)].$$

Proof:

$$\det[Tr_{L/K}(\nu_i \nu_j)] = \det\left[\sum_{\ell=1}^{n} f_\ell(\nu_i) f_\ell(\nu_j)\right]$$

$$= \det\left[f_\ell(\nu_i)\right] \cdot \det\left([f_k(\nu_j)]_{j,k}\right)^T = d_{L/K}(\nu_1, \cdots, \nu_n)$$

□

Theorem 12.6. Let L/K be a finite extension, where K is of characteristic zero. Let $\nu_1, \cdots, \nu_n$ be elements of L. Then

(i) The discriminant $d_{L/K}(\nu_1, \cdots \nu_n)$ lies in K and if all ν_i's lie in the integral closure of a domain R ($\subset K$) in L, then $d_{L/K}(\nu_1, \cdots, \nu_n) \in R$. In particular, if K is an algerbaic number field, then the discriminatnt $d_{K/\mathbf{Q}}$ of a set of algebraic integers is a rational integer.

(ii) If $\mu_i = \sum_{j=1}^{n} a_{ij}\nu_j \quad (i = 1, 2, ..., n; \; a_{ij} \in K)$, then $d_{L/K}(\mu_1, ..., \mu_n) = (\det[a_{ij}])^2 \; d_{L/K}(\nu_1, ..., \nu_n)$.

(iii) $d_{L/K}(\nu_1, \cdots, \nu_n) = 0$ if and only if the system $\nu_1, \cdots, \nu_n$ is linearly dependent over K.

(iv) If a generates L over K and $P(x) \in K[x]$ is its minimal polynominal, then

$$\begin{aligned} d_{L/K}(a) &= (-1)^m \det[a_{ij}] \\ &= (-1)^m \quad N_{L/K}(P'(a)) \end{aligned}$$

where $m = {}^nC_2 = \frac{n(n-1)}{2}$ and a_{ij}'s are defined by

$$a_i^j P'(a_i) = \sum_{i=0}^{n-1} a_{ij} \;\; a^i \qquad (j = 0, 1, \cdots, n-1),$$

$P'(x)$ is the formal derivative of $P(x)$ and $N_{L/K}(\alpha) = \Pi_{j=1}^{n} f_j(\alpha)$ is the norm mapping from L to K.

Proof: Part (i) follows from Theorem 12.5.

Part (ii) follows from the definition since $\det\,[f_j(\mu_i)] = \det[a_{ij}]\det[f_j(\nu_i)]$.

One part of (iii) is trivial. Because, if the elements $\nu_1, \cdots, \nu_n$ are linearly dependent over K, then for suitable $\alpha_1, \cdots, \alpha_n$ in K (not all of them 0) we have

$$\sum_{i=1}^{n} \alpha_i f_j(\nu_i) = 0, \quad \text{for } j = 1, \cdots, n$$

and therefore $d_{L/K}(\nu_1, \cdots, \nu_n) = 0$.

On the other hand, if $d_{L/K}(\nu_1, \cdots, \nu_n)$ vanishes, then the system

$$\sum_{i=1}^{n} x_i \; Tr_{L/K}(\nu_i\nu_j) = 0, \quad (j = 1, 2, \cdots, n$$

has a non-zero solution say $\alpha_1, \cdots, \alpha_n$ where $X_i \in K \ \forall\, i$. If the system $\nu_1, \cdots, \nu_n$ were linearly independent over K, then $\alpha = \alpha_1\nu_1 + \ldots + \alpha_n\nu_n$ would be non-zero and $Tr_{L/K}(\alpha\nu_i) = 0$ would hold for $i = 1, 2, \ldots n$. Hence, for all $y \in L$ we would have $Tr_{L/K}(\alpha y) = 0$ and in particular, for $y = \frac{1}{\alpha}$ we would get

$$n = Tr_{L/K}(1) = Tr_{L/K}(\alpha \cdot \frac{1}{\alpha}) = 0,$$

which is absurd.

Finally, to prove (iv), let us denote by $a_1, \cdots, a_n$ the conjugates of $a = a_1$ and put $b_i = P'(a_i)$, all these elements lying in a fixed algebraic closure of K.

$$\begin{aligned} \text{Then,}\quad d_{L/K}(a) &= \Pi_{i<j}(a_i - a_j)^2 \\ &= (-1)^m \ \Pi_{i=1}^n \ \Pi_{i\neq j}(a_i - a_j) \\ &= (-1)^m \ \Pi_{i=1}^n \ P'(a_i) \\ &= (-1)^m \ b_1 \cdots b_n \\ &= (-1)^m \ N_{L/K} \ (P'(a)). \end{aligned}$$

$$\begin{aligned} \text{Moreover,}\quad b_1 \cdots b_n \ \det[a_i^j] &= \det[a_i^j b_i] \\ &= \det \left[\sum_{k=1}^n a_{kj} a_k^i\right] \\ &= \det[a_{kj}] \ \det \ [a_j^k]. \end{aligned}$$

Since a generates L over K, from (ii) $\det[a_j^k] \neq 0$ and hence $b_1 b_2 \cdots b_n = \det\,[a_{ij}]$ and therefore $d_{L/K}(a) = (-1)^m \det[a_{ij}]$ □

Now, we leave the abstract situation and return to the algebraic number fields. Let K be a number field of degree n. If M is a free $\mathbf{Z}$-module which is contained in the ring of integers $\mathbf{O}_K$, then the discriminant $d_{K/\mathbf{Q}}(M)$ is defined to be the discriminant of any free basis of M. In particular, if $M = \mathbf{O}_K$, then this discriminant is called the *(absolute) discriminant* of the field K and is denoted by $d(K)$. That is, the discriminant of K is the discriminant of *any* integral basis of K. We observe that if the sets $\{a_1, \cdots, a_n\}$ and $\{b_1, \cdots, b_n\}$ both generate freely the $\mathbf{Z}$-module $\mathbf{O}_K$, then

$b_i = \sum_{j=1}^{n} a_{ij}a_j$, where $[a_{ij}]$ has elements from $\mathbf{Z}$ and also has an inverse with elements in $\mathbf{Z}$. Therefore, $\det[a_{ij}] = \pm 1$ and the definition is independent of the choice of the integral basis.

Theorem 12.7. The discriminant $d(K)$ of a field is either congruent to 1 (mod 4) or divisible by 4.

Proof: We consider an integral basis $w_1, w_2, \cdots, w_n$ of K.

Let $w_i^{(j)}, i, j = 1, \cdots, n$, be all the conjugates of the w_i's.

Then, $\det[w_i^{(j)}]$

$$= \sum_{\begin{pmatrix} 1 \cdots n \\ \mu_1 \cdots \mu_n \end{pmatrix} \text{is even}} w_1^{(\mu_1)} \cdots w_n^{(\mu_n)} - \sum_{\begin{pmatrix} 1 \cdots n \\ \mu_1 \cdots \mu_n \end{pmatrix} \text{is odd}} w_1^{(\mu_1)} \cdots w_n^{(\mu_n)}$$

$= A - B$, say.

If K_o is the minimal normal extension of $\mathbf{Q}$ which contain K, then A and B belong to K_0. Also, A + B and AB are invariant under the automorphism of K_0. Hence, A + B and AB are rational integers.

$$\begin{aligned} \text{Therefore, } d(K) &= (A-B)^2 \\ &\equiv (A+B)^2 - 4AB \\ &\equiv (A+B)^2 \pmod{4} \\ &\equiv 0 \text{ or } 1 \pmod{4}. \end{aligned}$$

□

12.3 Some results in geometry of numbers

Definition. For each $\mathbf{Z}$-basis $e = (e_1, ...e_n)$ of a lattice H in $\mathbf{R}^n$, let P_e be the half open parallelotope

$$P_e = \{x \in \mathbf{R}^n | x = \sum_{i=1}^{n} \alpha_i e_i, 0 \leq \alpha_i < 1\}.$$

Every point of $\mathbf{R}^n$ is congruent modulo H to one and only one point of P_e. We call P_e a *fundamental domain* for H. The Lebesgue measure $\mu(P_e)$

is called *the volume* of the lattice by abuse of language; we shall denote it by $\nu(H)$.

Since for another base $f = (f_1, ... f_n)$ of H, $f_i = \sum_{j=1}^{r} \alpha_{ij} e_j$ with $\alpha_{ij} \in \mathbf{Z}$, where (α_{ij}) is invertible with an integer inverse matrix, and hence satisfies $\mid \det(\alpha_{ij}) \mid = 1$, we have, $\mu(P_f) = \mid \det(\alpha_{ij}) \mid \mu(P_e) = \mu(P_e)$ and hence, $\mu(P_e)$ is independent of the choice of the base of H.

Theorem 12.8. (**Blichfeldt**) Let H be a lattice in $\mathbf{R}^n$ and let S be a measurable subset of $\mathbf{R}^n$ such that $\mu(S) > \nu(H)$. Then, there are $a, b \in S$, $a \neq b$ such that $a - b \in H$.

Proof: Let P_e be the fundamental parallelotope for H corresponding to the $\mathbf{Z}$ base $e = (e_1, \cdots, e_n)$ of H. Now, $S = \cup_{h \in H}(S \cap (h + P_e))$ and hence
$\mu(S) = \sum_{h \in H} \mu(S \cap (h + P_e))$ (Since μ is additive)
$= \sum_{h \in H} \mu((-h + S) \cap P_e)$. (Since μ is translation invariant)

If $(-h + S) \cap P_e$ $(h \in H)$ are pairwise disjoint, then

$$\mu(P_e) \geq \sum_{h \in H} \mu((-h + S) \cap P_e) = \mu(S),$$

contradicting the assumption.

Therefore, there are $h, h' \in H$, $h \neq h'$ such that

$$P_e \cap (-h + S) \cap (-h' + S) \neq \emptyset.$$

Therefore, $-h + a = -h' + b$ for some $a, b \in S$, which means $a - b = h - h'(\neq 0) \in H$. □

Corollary 12.8.1. (**Minkowski**) Let H be a lattice in $\mathbf{R}^n$ and let S be a measurable subset of $\mathbf{R}^n$ which is symmetric with respect to 0 and convex.

If $\mu(S) > 2^n \nu(H)$, then $S \cap (H - \{0\}) \neq \emptyset$.

If, in addition, S is compact, then one can replace $>$ by the weaker $\geq$ in the previous line.

Proof: Let $S' = \frac{1}{2}S$. Then, $\mu(S') = 2^{-n}\mu(S) > \nu(H)$. and by the above theorem, there are $y, z \in S'$ such that

$$y \neq z \text{ and } y - z \in H. \tag{12.1}$$

Now,

$$y - z = \frac{2y + (-2z)}{2}. \tag{12.2}$$

Here, $2y,\ 2z \in S$ and since S is symmetric about 0 we have $-2z \in S$.

Therefore, since S is convex, by (12.2), $y - z \in S$. This alongwith (12.1), implies that

$$y - z \in S \cap (H - \{0\}).$$

If S is compact we apply the above argument to $(1+\varepsilon)S$ for some $\varepsilon > 0$. Now, $(H - \{0\}) \cap (1 + \varepsilon)S$ is non empty and compact.

Consider the intersection $L = \cap_{\varepsilon_i > 0}(H - \{0\}) \cap (1 + \varepsilon_i)S$, $\varepsilon_i \to 0$ of non-empty nested compact sets. Then, $L \neq \emptyset$. Since S is compact, the result follows. □

Exercise 12.3. Let m be any odd positive integer. From Exercise 1.10 we know that there are positive integers a and b such that $a^2 + b^2 + 1 \equiv 0 \pmod m$. Consider the lattice H in $\mathbf{R}^4$ with the vectors $v_1 = (m, 0, 0, 0)$, $v_2 = (0, m, 0, 0)$, $v_3 = (a, b, 1, 0)$ and $v_4 = (b, -a, 0, 1)$ forming its lattice basis.

From the above result of Minkowski, show that the open ball $\mathcal{B}$ in $\mathbf{R}^4$ of radius $\sqrt{2m}$ and centre at the origin contains a non-zero point of H. Deduce that m is a sum of four square integers. Hence conclude that any positive integer is a sum of four squares. The last statement is the celebrated **Lagrange's four-square theorem**.

12.4 An estimation

In this section, we establish the existence of integral ideals, of norms bounded by specific bounds depending only on the particular number field, in every ideal class. We make use of Minkowski's result proved in the previous section.

Let K, r_1, r_2, and σ_i's are as defined next to Theorem 12.4.

Theorem 12.9. Let M be a free $\mathbf{Z}$-submodule of K of rank n and $(x_i)_{1 \le i \le n}$ a $\mathbf{Z}$-basis of M. Let $k = r_1 + r_2$. Let $\sigma_1, \cdots, \sigma_{r_1}, \sigma_{r_1+1}, \cdots, \sigma_k$ be the embeddings of K into $\mathbf{C}$; $\sigma_1, \cdots, \sigma_{r_1}$ are the real embeddings and $\sigma_{r_1+1}, \cdots, \sigma_k$

are complex embeddings where from each pair of conjugate ones only one is chosen. Now, we consider the mapping $\sigma : K \to \mathbf{R}^n$ defined by

$$\sigma(x) = (\sigma_1(x), \cdots, \sigma_{r_1}(x), \text{re } \sigma_{r_1+1}(x), \text{im } \sigma_{r_1+1}(x), \cdots, \text{re } \sigma_k(x), \text{im } \sigma_k(x))$$

where re z and im z denote real and imaginary parts of z respectively.

Then $\sigma(M)$ is a lattice in $\mathbf{R}^n$, and the volume of this lattice is

$$\nu(\sigma(M)) = 2^{-r_2} \mid \det(\sigma_i(x_j)) \mid . \tag{12.3}$$

Proof: We calculate the determinant D of the matrix whose i-th column is $(\sigma_1(x_i), \cdots, \sigma_{r_1}(x_i), \text{re } \sigma_{r_1+1}(x_i), \text{im } \sigma_{r_1+1}(x_i), \cdots, \text{re } \sigma_k(x_i), \text{im } \sigma_k(x_i))$. We know that re $z = \frac{1}{2}(z + \bar{z})$ and im $z = \frac{1}{2i}(z - \bar{z})$ for $z \in \mathbf{C}$. Remembering this, we take the factors $\frac{1}{2i}$ and $\frac{1}{2}$ out and add (r_1+2)-th row to the previous one etc. to obtain that $D = (2/i)^{-r_2} \det[\sigma_j(x_i)]$. From Theorem 12.6 (iii), it follows that $\det[\sigma_j(x_i)] \neq 0$. Hence $D \neq 0$ and therefore the vectors $\sigma(x_i)$'s are linearly independent in $\mathbf{R}^n$. This shows that $\sigma(M)$ is a lattice in $\mathbf{R}^n$ and the way we have calculated D, (12.3) does give the volume of this lattice. □

Corollary 12.9.1. If $d(K)$ is the absolute discriminant of a number field K, then $\nu(\sigma(\mathbf{O}_K)) = 2^{-r_2} \mid d(K) \mid^{1/2}$ and if $\mathtt{A}$ is an integral ideal $\nu(\sigma(\mathtt{A})) = 2^{-r_2} \mid d(K) \mid^{1/2} N(\mathtt{A})$.

Proof: We know that $\mathbf{O}_K$ and $\mathtt{A}$ are free $\mathbf{Z}$-modules of rank n. Therefore we can apply Theorem 12.9. If (x_i) is a $\mathbf{Z}$-base for $\mathbf{O}_K$, then $d(K) = (\det(\sigma_i(x_j)))^2$ and hence the first part of the corollary. For the second part we have only to notice that $\sigma(\mathtt{A})$ is a subgroup of $\sigma(\mathbf{O}_K)$ of index $N(\mathtt{A})$ and therefore a fundamental domain for $\sigma(\mathtt{A})$ can be constructed as the disjoint union of $N(\mathtt{A})$ copies of a fundamental domain for $\sigma(\mathbf{O}_K)$. □

Theorem 12.10. Let K be a field of degree n with signature $[r_1, r_2]$ and $\mathbf{O}_K$ its ring of integers. Let $\mathtt{A}$ be an ideal of $\mathbf{O}_K$ with $N(\mathtt{A}) = m$. Then there is a non-zero element a in $\mathtt{A}$ such that

$$\mid N_{K/\mathbf{Q}}(a) \mid \leq \left(\frac{4}{\pi}\right)^{r_2} \frac{n!}{n^n} \mid d(K) \mid^{1/2} m.$$

Proof: Let k and σ be as in the statement of Theorem 12.9. By Corollary 12.9.1, the image of $\mathtt{A}$ under σ is a lattice whose volume is

$$2^{-r_2} \mid d(K) \mid^{1/2} N(\mathtt{A}) = 2^{-r_2} \mid d(K) \mid^{1/2} m.$$

For $t > 0$, let $X_t \subset \mathbf{R}^n$ be defined by X_t

$$= \left\{ (x_1, \cdots, x_{r_1}, y_{r_1+1}, z_{r_1+1}, \cdots, y_k, z_k) : \sum_{i=1}^{r_1} | x_i | + 2 \sum_{j=1+r_1}^{k} (y_j^2 + z_j^2)^{1/2} < t \right\}.$$

We observe that X_t is convex, bounded and symmetric about the origin.

We claim that its volume is

$$V(X_t) = 2^{r_1} \left(\frac{\pi}{2}\right)^{r_2} \frac{t^n}{n!}.$$

If $r_1 = 1$, $r_2 = 0$, then $V(X_t) = 2t$ and if $r_1 = 0$, $r_2 = 1$, then $V(X_t) = \pi t^2/4$.

Hence our claim is true in these cases. Now, we proceed by induction. Assume that the result holds for $r_1 = A, r_2 = B$.

Then for $r_1 = A + 1, r_2 = B$, we have

$$\begin{aligned} V(X_t) &= \frac{2^{A-B}\pi^B}{(A+2B)!} 2 \int_0^t (t - x_{A+1})^{A+2B} dx_{A+1} \\ &= \frac{2^{A+1-B}\pi^B}{(A+2B+1)!} t^{A+2B+1} \end{aligned}$$

Similarly, for $r_1 = A$, $r_2 = B + 1$, we have

$$\begin{aligned} V(X_t) &= \frac{2^{A-B}\pi^B}{(A+2B)!} \int \int_{y^2+z^2 \leq t^2/4} (t - 2\sqrt{y^2 + z^2})^{A+2B} dydz \\ &= \frac{2^{A+1-B}\pi^{1+B}}{(A+2B+2)!} t^{A+2B+2} \end{aligned}$$

This proves the claim.

Now, given $\epsilon > 0$, let

$$t^n = (t(\epsilon))^n = \left(\frac{4}{\pi}\right)^{r_2} n! \mid d(K) \mid^{1/2} m + \epsilon.$$

Then,

$$V(X_t) = 2^{r_1} \left(\frac{\pi}{2}\right)^{r_2} \frac{t^n}{n!} > (2^{-r_2} \mid d(K) \mid^{1/2} m) 2^n.$$

Therefore, by Corollary 12.8.1, there exists $a \neq 0$ in $\mathtt{A}$ such that $\sigma(a) = (x_1, \cdots, x_{r_1}, y_{r_1+1}, z_{r_1+1}, \cdots, y_k, z_k)$ satisfies

$$\sum_{i=1}^{r_1} | x_i | + 2 \sum_{j=1+r_1}^{k} (y_j^2 + z_j^2)^{1/2} < t.$$

Now, using the above inequality and the inequality between arithmetic and geometric means,

$$| N_K(a) |^{1/n} = \left(\prod_{i=1}^{r_1} | x_i | \prod_{j=1+r_1}^{r_2} | y_j^2 + z_j^2 |^{1/2} \right)^{1/n} \leq \frac{t}{n}.$$

Therefore,

$$| N_K(a) | \leq \frac{t^n}{n^n}$$

which implies,

$$| N_K(a) | \leq \left(\frac{4}{\pi}\right)^{r_2} \frac{n!}{n^n} | d(K) |^{1/2} m + \frac{\epsilon}{n^n}.$$

Since $\epsilon > 0$ is arbitrary, and $\sigma(\mathtt{A}) \cap X_t$ is finite, the result follows. □

Corollary 12.10.1. In every ideal class there is an integral ideal of norm not exceeding $(\frac{4}{\pi})^{r_2} \frac{n!}{n^n} \mid d(K) \mid^{1/2}$.

Proof: The proof is the same as that of Theorem 12.4. □

Exercise 12.4. Only a finite number of algebraic number fields can have the same absolute discriminant.

Exercise 12.5. Find the class number of $\mathbf{Q}(\sqrt{-5})$.

12.5 Dirichlet's unit theorem

By abuse of language, the units in the ring of integers $\mathbf{O}_K$ of a number field K are called *units* of K. The multiplicative group of units of K will be denoted by $\mathbf{O}_K^*$.

Exercise 12.6. Let K be a number field. Then $u \in \mathbf{O}_K$ is a unit if and only if u is of norm ± 1.

The following important result of Dirichlet gives the structure of the group $\mathbf{O}_K^*$.

Theorem 12.11. **(Dirichlet's unit theorem)** Let K be an algebraic number field of degree n and signature $[r_1, r_2]$. Set $r = r_1 + r_2 - 1$. Then the group $\mathbf{O}_K^*$ of units of K is isomorphic to $G_K \times \mathbf{Z}^r$ where G_K is a finite cyclic group consisting of the roots of unity in K.

Proof. Let $\sigma_1, \cdots, \sigma_{r_1}, \sigma_{r_1+1}, \cdots, \sigma_{r_1+r_2}$ be the embeddings of K into $\mathbf{C}$; where as before $\sigma_1, \cdots \sigma_{r_1}$ are the real embeddings and $\sigma_{r_1+1}, \cdots \sigma_{r_1+r_2}$ are complex embeddings where from each pair of conjugate ones only one is chosen.

Consider the mapping $\mathcal{L} : K^* \to \mathbf{R}^{r_1+r_2}$ defined by

$$\mathcal{L}(x) = (\log \mid \sigma_1(x) \mid, \cdots, \log \mid \sigma_{r_1+r_2}(x) \mid).$$

This is a homomorphism and is called the *logarithmic embedding* of K^*.

Let B be a compact subset of $\mathbf{R}^{r_1+r_2}$. Let $A = \mathcal{L}^{-1}(B) \cap \mathbf{O}_K^*$.

Since B is bounded, there is a real number $\alpha > 1$ such that for all $x \in A, \alpha^{-1} \leq \mid \sigma_i(x) \mid \leq \alpha$ for all $i, 1 \leq i \leq n$. Therefore the elementary symmetric functions of the $\sigma_i(x)$'s are bounded in absolute value. Since they belong to $\mathbf{Z}$, the set of possible values for the symmetric functions of the $\sigma_i(x)$'s is a finite set. Therefore, the minimal polynomials for elements $x \in A$ are finitely many which in turn implies that the set A is finite.

This implies that the image $\mathcal{L}(\mathbf{O}_K^*)$ is a discrete subgroup of $\mathbf{R}^{r_1+r_2}$. Then by Theorem 8.6, $\mathcal{L}(\mathbf{O}_K^*)$ is a free $\mathbf{Z}$-module of rank $s \leq r_1 + r_2$. Since $\mathcal{L}(\mathbf{O}_K^*)$ is free, by Exercise 8.5, $\mathbf{O}_K^*$ is isomorphic to $G_K \times \mathcal{L}(\mathbf{O}_K^*) = G_K \times \mathbf{Z}^s$ where G_K is the kernel of the restriction of $\mathcal{L}$ to $\mathbf{O}_K^*$. The finiteness of A in the above paragraph implies that G_K is a finite group. Therefore G_K consists of roots of unity and is cyclic. Now, the roots of unity in K are algebraic integers and for a root of unity $x \in K$, $\mid \sigma_i(x) \mid = 1$ for all i. Therefore, G_K contains all the roots of unity in K.

It now remains to show that $s = r_1 + r_2 - 1$. If $x \in \mathbf{O}_K^*$, we have

$$\pm 1 = N(x) = \prod_{i=1}^{n} \sigma_i(x) = \prod_{i=1}^{r_1} \sigma_i(x) \prod_{i=r_1+1}^{r_1+r_2} \sigma_i(x)\overline{\sigma}_i(x)$$

which implies that the vector $\mathcal{L}(x) = (y_1, \cdots, y_{r_1+r_2})$ lies in a hyperplane $W \subset \mathbf{R}^{r_1+r_2}$ defined by the equation

$$\sum_{i=1}^{r_1} y_i + 2 \sum_{i=r_1+1}^{r_1+r_2} y_i = 0.$$

Thus $\mathcal{L}(\mathbf{O}_K^*)$ is a discrete subgroup of W and therefore $s \le r_1 + r_2 - 1$.

If we can now show that for any linear form $f \neq 0$ on W, $\exists\ u \in \mathbf{O}_K^*$ such that $f(\mathcal{L}(u)) \neq 0$, it will follow that $\mathcal{L}(\mathbf{O}_K^*)$ contains $r = r_1 + r_2 - 1$ linearly independent vectors and the proof will be complete. The projection $(y_1, y_2, \cdots, y_{r+1}) \mapsto (y_1, \cdots, y_r)$ from W on $\mathbf{R}^r$ being an isomorphism, for $y = (y_1, y_2, \cdots, y_{r+1}) \in W$, we can write

$$f(y) = c_1 y_1 + \cdots + c_r y_r, \quad c_i \in \mathbf{R}. \tag{12.4}$$

We fix a real number

$$\alpha \ge 2^n (1/2\pi)^{r_2} \mid d(K) \mid^{1/2}. \tag{12.5}$$

Now, given a r-tuple $a = (a_1, \cdots, a_r)$ of positive real numbers we take $a_{r+1} > 0$ such that

$$\prod_{i=1}^{r_1} a_i \prod_{i=r_1+1}^{r_1+r_2} a_i^2 = \alpha.$$

Let $B \subset \mathbf{R}^{r_1} \times \mathbf{C}^{r_2}$ be defined by

$$B = \{(y_1, y_2, \cdots, y_{r_1}, z_1, \cdots, z_{r_2}) : y_i \in \mathbf{R}, z_j \in \mathbf{C}, \mid y_i \mid \le a_i, \mid z_j \mid \le a_{j+r_1}\}.$$

Then B is compact, convex, symmetric about 0 and of volume

$$\prod_{i=1}^{r_1} 2a_i \prod_{i=r_1+1}^{r_1+r_2} \pi a_i^2 = 2^{r_1}\pi^{r_2}\alpha \ge 2^{n-r_2} \mid d(K) \mid^{1/2} \quad (\text{by}(12.5)).$$

Therefore by Corollary 12.8.1 and Corollary 12.9.1, there exists $x_a (\neq 0) \in \mathbf{O}_K$ such that $\mid \sigma_i(x_a) \mid \le a_i$ for $i = 1, \cdots, n$. Since x_a is an integer,

$$1 \le \mid N(x_a) \mid = \prod_{i=1}^{n} \mid \sigma_i(x_a) \mid \le \prod_{i=1}^{r_1} a_i \prod_{i=r_1+1}^{r_1+r_2} a_i^2 = \alpha. \tag{12.6}$$

Now, for any i,

$$\mid \sigma_i(x_a) \mid = \mid N(x_a) \mid \prod_{j \neq i} \mid \sigma_i(x_a) \mid^{-1} \ge \prod_{j\neq i} a_i^{-1} = a_i \alpha^{-1}.$$

Therefore, $a_i\alpha^{-1} \le \mid \sigma_i(x_a) \mid \le a_i$ for any i and hence

$$0 \le \log a_i - \log \mid \sigma_i(x_a) \mid \le \log \alpha.$$

Therefore from (12.4),

$$| f(\mathcal{L}(x_a)) - \sum_{i=1}^{r} c_i \log a_i | \leq (\sum_{i=1}^{r} | c_i |) \log \alpha. \tag{12.7}$$

We choose $\beta > (\sum_{i=1}^{r} | c_i |) \log \alpha$.

Now, for every positive integer h, we select r positive real numbers $a_{i,h}$ $(i = 1, \cdots, r)$ satisfying $\sum_{i=1}^{r} c_i \log a_{i,h} = 2\beta h$. Let $x_h = x_{a(h)}$ be the algebraic integer corresponding to the r-tuple $a(h) = (a_{1,h}, \cdots, a_{r,h})$.

Then from (12.7), $| f(\mathcal{L}(x_h)) - 2\beta h | < \beta$, which implies

$$(2h-1)\beta < f(\mathcal{L}(x_h)) < (2h+1)\beta. \tag{12.8}$$

It follows from (12.8) that the numbers $f(\mathcal{L}(x_h))$ are all distinct. Again, from (12.6), $| N(x_h) | \leq \alpha$, and hence by Exercise 12.2, there are only finitely many distinct ideals of the form $x_h\mathbf{O}_K$. Therefore there are two distinct indices h and k such that $x_h\mathbf{O}_K = \mathbf{O}_K x_k$ which implies that $x_h = ux_k$ for some unit u. Since f is linear, $f(\mathcal{L}(u)) = f(\mathcal{L}(x_h)) - f(\mathcal{L}(x_k)) \neq 0$. □

Exercise set D

D.1 For any two integral ideals $\mathtt{A}, \mathtt{B}$ in an algebraic number field K, there exists $\omega \in \mathbf{O}_K$ such that the greatest common divisor $(\mathtt{AB}, (\omega))$ is $\mathtt{A}$.

D.2 Given any integral ideal $\mathtt{A}$ in an algebraic number field K, there exist $\alpha, \omega \in \mathbf{O}_K$ such that $\mathtt{A} = \omega\mathbf{O}_K + \alpha\mathbf{O}_K$. That is, any integral ideal can be generated, over $\mathbf{O}_K$, by two algebraic integers in K.

D.3 For a rational prime $p > 0$, let $\zeta_p = e^{\frac{2\pi i}{p}}$ and $\mathbf{O}_K$ be the ring of integers of $K = \mathbf{Q}[\zeta_p]$. Show that

(i) $(1 - \zeta_p)\mathbf{O}_K \cap \mathbf{Z} = p\mathbf{Z}$.

(ii) $Tr(y(1 - \zeta_p)) \in p\mathbf{Z}$ for any $y \in \mathbf{O}_K$.

(iii) If $\alpha = a_0 + a_1\zeta_p + \cdots + a_{p-2}\zeta_p^{p-2}$ with a_i's in $\mathbf{Q}$ is an element of $\mathbf{O}_K$, then show that all the a_i's are in $\mathbf{Z}$, which amounts to showing that $\mathbf{O}_K = \mathbf{Z}[\zeta_p]$.

D.4 Let m be a positive integer and ζ_m denote the number $e^{\frac{2\pi i}{m}}$. Also, let D denote the ring of integers of $K = \mathbf{Q}[\zeta_m]$ and $f(x) \in \mathbf{Z}[x]$ the monic irreducible polynomial of ζ_m. Then, show that

(i) For a rational prime $p > 0$ such that $p \not| m$, ζ_m^p is also a root of $f(x)$.

(ii)

$$f(x) = \prod_{\substack{(a,m)=1 \\ 1 \leq a < m}} (x - \zeta_m^a).$$

(iii) For a rational prime $p > 0$ such that $p \not| m$ and $a \in \mathbf{Z}$, $p | f(a) \iff$ the multiplicative order of a mod p is m.

(iv) For a rational prime $p > 0$ such that $p \nmid m$, $p | f(a)$ for some $a \in \mathbf{Z}$ $\iff$ $p \equiv 1 (\mathrm{mod}\ m)$.

Deduce that for integers $m \geq 1$, there are infinitely many primes $p \equiv 1 (\mathrm{mod}\ m)$.

Chapter 13

Quadratic fields

After gathering some knowledge about the general theory of number fields, in the present chapter, we look into one important class of these fields and have some explicit informations about its members.

13.1 Integral bases and discriminants

Definition. If K is a number field such that $[K : \mathbf{Q}] = 2$, then K is called a *quadratic field.* It follows that a quadratic number field is of the form $\mathbf{Q}[\sqrt{d}]$ (consisting of complex numbers $a + b\sqrt{d}, a, b \in \mathbf{Q}$) where d is a fixed integer, positive or negative, which is not a rational square.

A quadratic field K is called a *real* or an *imaginary quadratic field* according as $K \subset \mathbf{R}$ or not.

We note that a quadratic field K is real if and only if $K = \mathbf{Q}(\sqrt{m})$, where m belonging to $\mathbf{Z}$ is a square-free integer greater than 1. If K is an imaginary quadratic field, then $K \cap \mathbf{R} = \mathbf{Q}$.

Remark 13.1. Let $K = \mathbf{Q}(\sqrt{m})$, (m square-free $\in \mathbf{Z}$) be a quadratic field. Let $\alpha \in \mathbf{O}_K$. Then $\alpha = p + q\sqrt{m}$ for some $p, q \in \mathbf{Q}$. Now, $Tr_K(\alpha) = 2p$ and $N_K(\alpha) = p^2 - q^2m$ are both in $\mathbf{Z}$. Writing $a = 2p$ and $b = p^2 - q^2m$, we have $(a^2 - 4q^2m)/4 = b \in \mathbf{Z}$. In other words,

$$a^2 - 4q^2m \equiv 0 \pmod 4. \tag{13.1}$$

Now, $a^2 - 4q^2m \in \mathbf{Z}$ and $a \in \mathbf{Z}$, imply that $4q^2m \in \mathbf{Z}$. Now, m being square-free, if we write $q = s/l$, where $s, l \in \mathbf{Z}$ with $(s, l) = 1$, then l can be 1 or 2. In other words, $q = f/2$ with $f \in \mathbf{Z}$.

So,

$$\alpha = \frac{a}{2} + \frac{f}{2}\sqrt{m}, \quad \text{with } a, f \in \mathbf{Z}.$$

We want to find an integral basis for $\mathbf{O}_K$.

We consider the two cases:

Case-I ($m \equiv 1 \pmod 4$)

In this case, from (13.1) we get that $a^2 \equiv f^2 \pmod 4$.

Therefore, a and f are both even or both odd.

In either case

$$\alpha = u + v\frac{1 + \sqrt{m}}{2}, \quad \text{with } u, v \in \mathbf{Z}.$$

Case-II ($m \equiv 2, 3 \pmod 4$)

In this case, (13.1) implies that both a and f are even.

Hence

$$\alpha = u' + v'\sqrt{m} \text{ with } u', v' \in \mathbf{Z}.$$

To sum up,

$$\mathbf{O}_K = \begin{cases} \mathbf{Z} + \mathbf{Z}\frac{1+\sqrt{m}}{2} & \text{for } m \equiv 1 \pmod 4 \\ \mathbf{Z} + \mathbf{Z}\sqrt{m} & \text{for } m \equiv 2, 3 \pmod 4 \end{cases} \tag{13.2}$$

We observe that m being square-free, the case $m \equiv 0 \pmod 4$ does not arise.

Remark 13.2. Let $K = \mathbf{Q}(\sqrt{m})$ be a quadratic field where $m \in \mathbf{Z}$ is square-free. With the help of the informations regarding integral bases as obtained in (13.2) we now calculate the discriminant $d(K)$ for quadratic number fields.

If $m \equiv 1 \pmod 4$, then

$$d = d(\mathbf{Q}(\sqrt{m})) = \begin{vmatrix} 1 & \frac{1+\sqrt{m}}{2} \\ 1 & \frac{1-\sqrt{m}}{2} \end{vmatrix}^2 = m.$$

If $m \equiv 2, 3 \pmod 4$, then a similar calculation yields $d = 4m$.

From (13.2) and Remark 13.2, we see that in all cases,

$$\mathbf{O}_K = \mathbf{Z} + \mathbf{Z}\frac{d+\sqrt{d}}{2}.$$

Therefore, we have the following theorem:

Theorem 13.1. The discriminant uniquely determines a quadratic field.

13.2 Splitting of rational primes

Definition. Let K be a quadratic field and $\wp$ be any prime ideal in $\mathbf{O}_K$. Now, $\wp$ contains a prime number $p \in \mathbf{Z}, p > 0$ and $\wp$ occurs in the factorization $\wp_1 \cdots \wp_r$ of $p\mathbf{O}_K$ into prime ideals in $\mathbf{O}_K$. Therefore, $p^2 = N_K(p) = N(p\mathbf{O}_K) = N(\wp_1)\cdots N(\wp_r)$, from which we conclude that $N(\wp) = p$ or p^2.

Hence, writing $\wp'$ for the image of $\wp$ under the nontrivial automorphism of K, one of the following cases will arise:

i) $p\mathbf{O}_K = \wp\wp', \wp \neq \wp'$.

ii) $p\mathbf{O}_K = \wp = \wp'$.

iii) $p\mathbf{O}_K = \wp^2, \wp = \wp'$.

In case i), we say that p *splits* in K. In case ii), we say that p *stays prime* in K. Finally if iii) occurs, then we say that p is *ramified* in K.

Theorem 13.2. Let K be a quadratic field with discriminant d as above. Then for an odd prime $p \in \mathbf{Z}$,

i) p splits in K if and only if $\left(\frac{d}{p}\right) = 1$.

ii) p is ramified in K if and only if $\left(\frac{d}{p}\right) = 0$.

iii) p stays prime in K if and only if $\left(\frac{d}{p}\right) = -1$.

Proof:

i) Let $\left(\frac{d}{p}\right) = 1$. Then there is $y \in \mathbf{Z}$ satisfying

$$y^2 \equiv d \pmod{p}. \tag{13.3}$$

Let $\wp$ be the ideal generated by p and $y + \sqrt{d}$.

Then

$$\wp\wp' = (p^2, p(y+\sqrt{d}), p(y-\sqrt{d}), y^2 - d)\mathbf{O}_K. \tag{13.4}$$

By (13.3), $p \mid (y^2 - d)$ and hence by (13.4)

$$\wp\wp' \subset p\mathbf{O}_K. \tag{13.5}$$

Now, the fact that the elements $p(y+\sqrt{d}), p(y-\sqrt{d}) \in \wp\wp'$ implies that $2py \in \wp\wp'$. Therefore, observing that $p = \gcd(p^2, 2py) = l_1p^2 + l_2 2py$ for some integers l_1, l_2, we see that $p \in \wp\wp'$ and hence $p\mathbf{O}_K \subset \wp\wp'$. This together with (13.5) implies that $p\mathbf{O}_K = \wp\wp'$. Since at most two prime ideals of $\mathbf{O}_K$ can divide $p\mathbf{O}_K$, $\wp$ and $\wp'$ are prime ideals. Also $p \in \wp$ and $2d = \sqrt{d}((y+\sqrt{d}) - (y-\sqrt{d})) \in \wp + \wp'$. Therefore, $1 = \gcd(p, 2d) \in \wp + \wp'$ which implies that $\wp \neq \wp'$.

Conversely, let $p\mathbf{O}_K = \wp\wp'$ where $\wp$ is a prime ideal and $\wp \neq \wp'$. Then $N(\wp) = N(\wp') = p$. There is $\alpha \in \wp$ such that $\alpha \notin p\mathbf{O}_K = \wp\wp'$. This implies that

$$\alpha = k + l\frac{d+\sqrt{d}}{2}, \tag{13.6}$$

where $k, l \in \mathbf{Z}$ are such that $p \nmid (k, l)$.

Since $\alpha \in \wp$, $\alpha\mathbf{O}_K = \wp\wp''$ for some ideal $\wp''$ of $\mathbf{O}_K$. This implies that $p = N(\wp)$ divides $N(\alpha\mathbf{O}_K) = \mid N_K(\alpha) \mid = \left|\left(k + l\frac{d}{2}\right)^2 - l^2\frac{d}{4}\right|$.

Hence,

$$(2k + dl)^2 \equiv l^2 d \pmod{p}. \tag{13.7}$$

If $p \mid l$, then $p \mid (2k+dl)^2$, which further implies that $p \mid 2k$. Since p is odd, we have $p \mid k$. Thus $p \mid (k, l)$, which is a contradiction to (13.6). Therefore p does not divide l. From (13.7), we therefore have $s^2 \equiv d \pmod{p}$ for $s \in \mathbf{Z}$, that is, $\left(\frac{d}{p}\right) = 1$.

ii) Let $\left(\frac{d}{p}\right) = 0$. We consider the ideal $\wp = p\mathbf{O}_K + \sqrt{d}\mathbf{O}_K$.

Then $\wp^2 = (p^2, p\sqrt{d}, d)\mathbf{O}_K = p\mathbf{O}_K$, since $p = \gcd(d, p^2)$. Also $\wp$ is necessarily a prime ideal.

Conversely, if $p\mathbf{O}_K = \wp^2$ where $\wp$ is a prime ideal of $\mathbf{O}_K$, then there exists $\theta = k + l\frac{d+\sqrt{d}}{2} \in \wp$ where $k, l \in \mathbf{Z}$ such that $\theta \notin p\mathbf{O}_K$. Since $\theta^2 \in p\mathbf{O}_K$, observing that

$$\theta^2 = \frac{1}{4}((2k + ld)(2k - ld) + l^2 d) + l(2k + ld)\frac{d + \sqrt{d}}{2}$$

we get that

$$p \mid ((2k + ld)(2k - ld) + l^2 d) \tag{13.8}$$

and

$$p \mid l(2k + ld). \tag{13.9}$$

If p divides l, then from (13.8), p divides $(2k + ld)$ or $(2k - ld)$. Since p is odd, this would imply that $p \mid k$. But then $p \mid l$ and $p \mid k$ imply $\theta \in p\mathbf{O}_K$, which is a contradiction.

Therefore, p does not divide l and from (13.9), p divides $(2k + ld)$. But (13.8) would then imply $p \mid l^2 d \Rightarrow p \mid d \Rightarrow \left(\frac{d}{p}\right) = 0$.

iii) The validity of (iii) is a consequence of (i) and (ii).

□

Regarding ramification of the prime 2, the condition is described in the following theorem which we state without proof.

Theorem 13.3. Let K be a quadratic field with discriminant d. Then

i) 2 splits in K if and only if $d \equiv 1 \pmod 8$.

ii) 2 is ramified in K if and only if $d \equiv 0 \pmod 4$.

iii) 2 stays prime in K if and only if $d \equiv 5 \pmod 8$.

13.3 The group of units

Now, we proceed to study the group of units in quadratic fields. First, we take up the case of imaginary quadratic fields.

Theorem 13.4. Let $K = \mathbf{Q}[\sqrt{-m}]$ be an imaginary quadratic field. Then the group of units in K is described as follows.

(i) If $m = 1$, then $\mathbf{O}_K^*$ is $\{\pm i, \pm 1\}$, that is, the group of fourth roots of unity.

(ii) If $m = 3$, then $\mathbf{O}_K^*$ is the group of sixth roots of unity.

(iii) If m is any positive integer other than 1 or 3, $\mathbf{O}_K^*$ is $\{\pm 1\}$.

Proof: With the notations of Theorem 12.11, for an imaginary quadratic field we have $r_1 = 0$ and $r_2 = 1$ and hence $r = r_1 + r_2 - 1 = 0$. Therefore, by Theorem 12.11, the group $\mathbf{O}_K^*$ of units of K is G_K where G_K is the finite cyclic group consisting of the roots of unity in K.

To obtain more precise informations, we proceed as follows. From (13.2),

$$\mathbf{O}_K = \begin{cases} \mathbf{Z} + \mathbf{Z}\frac{1+\sqrt{m}}{2} & \text{for } -m \equiv 1 \pmod 4 \\ \mathbf{Z} + \mathbf{Z}\sqrt{m} & \text{for } -m \equiv 2, 3 \pmod 4 \end{cases}$$

For $-m \equiv 2, 3 \pmod 4$, considering an element $\alpha = a + b\sqrt{-m}$ of $\mathbf{O}_K$, in order that α is a unit, by Exercise 12.6, we must have $a^2 + mb^2 = 1$. If $m > 1$, then b has to be zero and $a = \pm 1$. If $m = 1$, then the solutions are $a = 0, b = \pm 1$ and $a = \pm 1, b = 0$.

Proceeding similarly, in the case of $-m \equiv 1 \pmod 4$, except for $m = 3$, the units are $\{\pm 1\}$. And for $m = 3$, calculations show that they are as stated in (ii) in the theorem. □

If K is a real quadratic field, then $r_1 = 2$, $r_2 = 0$. Therefore $r_1 + r_2 - 1 = 1$. Also since $G_K = \{\pm 1\}$ in this case, from Theorem 12.11 we have:

Theorem 13.5. The group of units of a real quadratic field is isomorphic to $\{\pm 1\} \times \mathbf{Z}$.

Definition. From Theorem 13.5,we see that in the case of a real quadratic field K, every unit ϵ in K can be written in the form $\pm\epsilon_1^n, n \in \mathbf{Z}$ for a fixed unit ϵ_1 in K. Also $\epsilon_1 \neq \pm 1$. Here the role of ϵ_1 can be played by $\epsilon_1^{-1}, -\epsilon_1$ or

$-\epsilon_1^{-1}$ as well. But among $\epsilon_1, \epsilon_1^{-1}, -\epsilon_1, -\epsilon_1^{-1}$, exactly one of them is greater than 1. We call it the *fundamental unit* of K.

We now consider the following Diophantine equation, the so-called *Pell's equation:*

$$a^2 - mb^2 = \pm 1, \tag{13.10}$$

where $m \neq 0$ is a square-free integer.

We seek the integer solutions of (13.10).

If $m < 0$, these solutions are $(\pm 1, 0)$ for $m < -1$ and $(\pm 1, 0), (0, \pm 1)$ for $m = -1$.

However if $m > 1$, it is a nontrivial fact that (13.10) has infinitely many solutions in integers. We use our knowledge about units in real quadratic fields to draw precise conclusions regarding this.

Let $K = \mathbf{Q}(\sqrt{m})$, where $m > 1$ is a square-free integer.

Once again, we divide the analysis in two cases.

Case-I ($m \equiv 2$ or $3 \pmod 4$)

In this case

$$\mathbf{O}_K = \mathbf{Z} + \mathbf{Z}\sqrt{m}.$$

As the units of K are integers of norm ± 1, the units of K which are greater than 1, are the numbers $\alpha = a + b\sqrt{m}$ with $a, b \in \mathbf{Z}$ and $a > 0, b > 0$ such that

$$N(\alpha) = (a + b\sqrt{m})(a - b\sqrt{m}) = a^2 - mb^2 = \pm 1.$$

Therefore, it is clear that if we take the fundamental unit $a_1 + b_1\sqrt{m}$ of K and put

$$a_n + b_n\sqrt{m} = (a_1 + b_1\sqrt{m})^n, n \geq 1,$$

then the sequence (a_n, b_n) lists all the solutions of (13.10). If the fundamental unit is of norm one, the sequence (a_n, b_n) gives solutions only for the equation $a^2 - mb^2 = 1$; in this case the equation $a^2 - mb^2 = -1$ has no solutions in natural numbers. If the fundamental unit is of norm -1, the solutions of $a^2 - mb^2 = 1$ are given by the sequence (a_{2n}, b_{2n}) and those of $a^2 - mb^2 = -1$ by the sequence (a_{2n+1}, b_{2n+1}). For instance the first case occurs when $m = 3$, the second when $m = 2$.

Case-II ($m \equiv 1 \pmod 4$)

In this case $\mathbf{O}_K = \{\frac{1}{2}(a+b\sqrt{m})\}$ where $a, b \in \mathbf{Z}$ are of the same parity. If $\frac{1}{2}(a+b\sqrt{m})$ is a unit of K, then we must have $a^2 - mb^2 = \pm 4$ and solutions of this equation can be obtained as in the earlier case. In this case the solutions of (13.10) correspond to the units $a+b\sqrt{m}$ which belong to the ring $\mathbf{Z}[\sqrt{m}]$.

13.4 Norm-Euclidean number fields

Definition. We call an algebraic number field K *norm-Euclidean* if its ring of integers is Euclidean with the size function $\phi(x) = |N_{K/\mathbf{Q}}(x)|$.

Exercise 13.1. Show that a number field K is norm -Euclidean iff for every element $a \in K$ there exists an integer b of the same field with

$$|N_{K/\mathbf{Q}}(a-b)| < 1.$$

Exercise 13.2. An imaginary quadratic field $K = \mathbf{Q}(\sqrt{-m})$ is norm -Euclidean if and only if $m = 1, 2, 3, 7$ or 11.

Remark 13.3. Regarding unique factorizations in imaginary quadratic fields it was proved by Gauss that for negative integers m, the ring of integers of $\mathbf{Q}(\sqrt{m})$ is a UFD if $m = -1, -2, -3, -7, -11, -19, -43, -67$, or -163. He also conjectured that there are no others. More than 150 years after this conjecture, it was finally proved by Baker and Stark in 1966. It should be mentioned that in essence a proof had been given earlier by Heegner.

Remark 13.4. The corresponding Gauss conjecture for real quadratic fields, which says that there are infinitely many positive integers m such that $\mathbf{Q}(\sqrt{m})$ has class number 1 is still open.

Remark 13.5. In the real quadratic case, it is known that $\mathbf{Q}(\sqrt{m})$ is norm -Euclidean iff m has one of the sixteen values 2, 3, 5, 6, 7, 11, 13, 17, 19, 21, 29, 33, 37, 41, 57 and 73.

Remark 13.6. In the imaginary case, it is known that there are no others, than the ones listed in Exercise 13.2, which can possibly be Euclidean with any other size function. In view of Remark 13.3, this result supplies us with the examples of fields $\mathbf{Q}(\sqrt{-19})$, $\mathbf{Q}(\sqrt{-43})$, $\mathbf{Q}(\sqrt{-67})$, $\mathbf{Q}(\sqrt{-163})$ whose rings of integers are principal ideal domains but not Euclidean domains.

***Remark 13.7.** Assuming the Generalized Riemann Hypothesis, it was shown by Weinberger [We 1973] that for an algebraic number field containing infinitely many units, its ring of integers is a Euclidean domain if and only if it is a principal ideal domain. Very recently, Clark [Cl 1994] has shown that the ring of integers of the field $\mathbf{Q}(\sqrt{69})$ is Euclidean with some size function, as is expected by Weinberg's result. In this connection, one may look into the papers [GMM 1987] and [Le 1995].

Chapter 14

Solutions to selected exercises

0.1 Since in a field F, all non-zero elements are units, the first assertion is obvious. Now, let R be a ring which has no proper ideal. Let r be any non-zero element of R. Since (r), the principal ideal generated by r, is non-zero, by our assumption it must be R. This means that 1 belongs to (r). Hence, there is $r' \in R$ such that $rr' = 1$. That is, r' is the inverse of r. Therefore, every non-zero element of R is a unit, which means that R is a field.

0.2 Let D be a finite integral domain. Let $d_1, d_2, \cdots, d_n$ be the distinct elements of D. Let d be a non-zero element of D. We consider the elements $dd_1, dd_2, \cdots, dd_n$. Since D does not have any zero -divisor, for $i \neq j$, $dd_i \neq dd_j$. Therefore, the subset $\{dd_1, dd_2, \cdots, dd_n\}$ of D, having the same number of elements as D, is equal to D. Therefore, $dd_i = 1$ for some $i \in \{1, 2, \cdots, n\}$. Thus any non-zero element of R has an inverse and we are through.

0.3 Let $\mathcal{S}$ be the set of all ideals other than R of a non-zero ring R. Since the zero-ideal belongs to $\mathcal{S}$, $\mathcal{S}$ is non empty. We consider $\mathcal{S}$ with the partial ordering given by set inclusion. A maximal element of $\mathcal{S}$ in this partial ordering is a maximal ideal by definition. We use Zorn's lemma to show that $\mathcal{S}$ has a maximal element.

Given a totally ordered subset $\mathcal{F}$ of $\mathcal{S}$, we observe that the union of all elements of $\mathcal{F}$ is an element of $\mathcal{S}$ and clearly an upper bound of

$\mathcal{F}$. Therefore, by Zorn's lemma $\mathcal{S}$ has a maximal element M.

1.1 (a) We already observed that for any integer n, the set $n\mathbf{Z}=\{nr|r \in \mathbf{Z}\}$ is a subgroup of $(\mathbf{Z}, +)$.

Let H be any subgroup of $(\mathbf{Z}, +)$. If H does not contain any non-zero element, then $H = 0\mathbf{Z}$. If H contains a non-zero element a, then H will contain a positive integer (since either a or $-a$ will be positive). Let m be the smallest positive integer in H. Since H is a subgroup, $m\mathbf{Z} \subset H$. To see this, if k is a positive integer $mk = \underbrace{m + \cdots + m}_{k \text{ times}} \in H$. Again, $m(-k) = -mk \in H$ and $m0 = 0 \in H$. If possible, let $b \in H \setminus m\mathbf{Z}$. Now, by the division algorithm $b = qm+r$ where $0 \leq r < m$. Since b does not belong to $m\mathbf{Z}$, $r \neq 0$. But then $r = b - qm \in H$ – contradicting the minimality of m. Hence, $H = m\mathbf{Z}$. Thus the sugroups of $(\mathbf{Z}, +)$ are precisely the subsets $n\mathbf{Z}$ for $n \in \mathbf{Z}$.

(b) Let G be a cyclic group. Let a be a generator of G. Let $f : \mathbf{Z} \to G$ be the map defined by $f(n) = a^n$. Clearly this is a surjective homomorphism. The kernel of f is a subgroup of $\mathbf{Z}$ and hence it is $m\mathbf{Z}$ for some integer $m \geq 0$. If $m = 0$, then G is isomrphic to $\mathbf{Z}$ and if $m > 0$, then G is isomorphic to $\mathbf{Z}/m\mathbf{Z}$, by the first isomorphism theorem.

(c) Follows from part (a) above.

1.2 If p_1 $(= 3), p_2$ $(= 7), \cdots, p_r$ is a given sequence of primes of the form $4n + 3$ in increasing order, then we consider the number $m = 4p_2p_3 \cdots p_r + 3$. The primes dividing the odd integer m can not all be of the form $4n + 1$. That is, m must have a prime divisor of the form $4n + 3$. Clearly, it can not be any one among the p_i's, $i = 1, 2, \cdots, r$.

1.3 (a) There are mm' numbers of the form $a'm + am'$ where a and a' runs through complete sets of residues modulo m and m' respectively. If $a'm+am' \equiv b'm+bm' \pmod{mm'}$, then $a'm \equiv b'm \pmod{m'}$ and $am' \equiv bm' \pmod{m}$. Since $(m, m') = 1$, it follows that $a' \equiv b' \pmod{m'}$ and $a \equiv b \pmod{m}$.

(b) If a runs through a reduced residue system modulo m and a' runs through a reduced residue system modulo m', we show that $a'm + am'$ runs through a reduced residue system modulo mm'.

In part (a), we have already proved that the numbers $a'm + am'$ are incongruent. Let p be a prime which divides $(mm', a'm + am')$. Then p divides m or m'. If $p|m$, then p has to divide am'. Since, $(m, m') = 1$, it follows that $p|a$. Thus $p|(a, m)$. This contradicts the fact that a runs through a reduced residue system modulo m. Therefore, $(mm', a'm + am') = 1$.

Let d be any integer such that $(d, mm') = 1$. By part (a), there are integers a and a' such that $d = a'm + am'$. Now, $1 = (d, m) = (a'm + am', m) = (am', m) = (a, m)$. Similarly, $(a', m') = 1$.

This completes part (b).

(c) Because of part (b), it is enough to prove $\phi(p^a) = p^a\left(1 - \frac{1}{p}\right)$, where p is a prime and a is a positive integer. Now, among the p^a integers $1, 2, \cdots, p^a$, exactly p^{a-1} integers are multiples of p. Therefore $\phi(p^a) = p^a - p^{a-1} = p^a\left(1 - \frac{1}{p}\right)$.

1.4 We consider the set $\{1, 2, \cdots, m\}$, which is a complete residue system modulo m. Now we observe that the set $\{a, 2a, \cdots, ma\}$ is also a complete residue system modulo m. For, $ia \equiv ja \pmod{m}$ implies $i \equiv j \pmod{m}$ since $(a, m) = 1$. Therefore, exactly one of the elements $a, 2a, \cdots, ma$ is congruent to b modulo m.

1.5 As in Exercise 1.4, we observe that given a reduced residue system $\{a_1, a_2, \cdots, a_{\phi(m)}\}$ modulo m, the set $\{aa_1, aa_2, \cdots, aa_{\phi(m)}\}$ is also a reduced residue system modulo m.

Therefore, $a_1 a_2 \cdots a_{\phi(m)} \equiv aa_1 aa_2 \cdots aa_{\phi(m)} \pmod{m}$.

Since each a_i is relatively prime to m, this implies

$1 \equiv a^{\phi(m)} \pmod{m}$.

1.6 Since p divides n^2+1, it follows that $(p,n) = 1$. Therefore, by Exercise 1.5,

$$n^{p-1} \equiv 1 \pmod{p}. \tag{14.1}$$

It is given that

$$n^2 \equiv -1 \pmod{p}. \tag{14.2}$$

If possible, let p be of the form $4m+3$. Then $\frac{p-1}{2}$ is odd and hence from (14.2) we have

$$n^{p-1} \equiv -1 \pmod{p}. \tag{14.3}$$

From (14.1) and (14.3) we have $1 \equiv -1 \pmod{p}$ which is impossible as p is odd.

Hence any odd prime dividing a number of the form n^2+1 must be of the form $4m+1$.

Now, if possible, let the only primes of the form $4m+1$ be

$$p_1 = 5, p_2 = 13, \cdots, p_r.$$

Now, consider the number $d = (2p_1p_2\cdots p_r)^2 + 1$.

Clearly a prime divisor of d is odd. Now, by the first part of this exercise, any prime divisor of d must be of the form $4m+1$. This leads to a contradiction as none of the primes $p_1, p_2, \cdots, p_r$ can divide d. Hence there must be infinitely many primes of the form $4m+1$.

1.7 For any integer n we see that n^2 is either 0 or 1 (mod 3). Therefore, n^2+1 is either 1 or 2 (mod 3). Therefore, the congruence $x^2+1-3y^5 = 0 \pmod{3}$ does not have any solution which implies that the given equation does not have solutions in integers.

1.10 One observes that the $(p+1)/2$ elements $\{a^2 : 0 \le a \le (p-1)/2\}$ are distinct modulo p. So are the elements $\{-a^2-1 : 0 \le a \le (p-1)/2\}$. Therefore by pigeonhole principle, there is an element of the first set which is equal to one element in the second set modulo p. This proves Part (a).

Observe that if the integers a and b are such that $a^2 + b^2 + 1 \equiv 0 \pmod{p^k}$, then at least one of a and b, say a, is not divisible by p. Now, $a^2 = -b^2 - 1 + qp^k$ for some integer q. Since a is not divisible by the odd prime p, we have gcd $(2a, p) = 1$ and hence $q = x2a + yp$ for some integers x, y.

Now, it is easy to see that $(a - xp^k)^2 \equiv -b^2 - 1 \pmod{p^{k+1}}$ and this shows that Part (b) follows by induction from Part (a).

Finally, if the prime factorization of m is $m = \prod_{i=1}^{l} p_i^{r_i}$ where p_i's are distinct primes, by getting hold of (by Part (b)) a_i, b_i satisfying $a_i^2 + b_i^2 + 1 \equiv 0 \pmod{p_i^{r_i}}$ for each i and then finding a, b (by Chinese remainder theorem) such that $a \equiv a_i \pmod{p_i^{r_i}}$ and $b \equiv b_i \pmod{p_i^{r_i}}$ we have $a^2 + b^2 + 1 \equiv 0 \pmod{m}$.

A.1 Let $n > 1$ be any integer. If possible, suppose $\frac{1}{2} + \frac{1}{3} + \cdots + \frac{1}{n} = m$, an integer.

Let r be the positive integer such that $2^r \leq n < 2^{r+1}$.

Writing $2^r = d$, we have

$$\sum_{\substack{2 \leq i \leq n \\ i \neq d}} \frac{1}{i} = m - \frac{1}{d}. \tag{14.4}$$

If L is the least common multiple of the denominators on the left-hand side, then $L = 2^{r-1} \times L_1$ where L_1 is odd.

But after adding the fractions on the left-hand side in (14.4), we have $\frac{N}{L} = \frac{md-1}{d}$ for some integer N. Now, $Nd = (md-1)L$ where the largest power of 2 dividing the right-hand side is 2^{r-1} (since $md - 1$ is odd) and that for the left-hand side is at least $d = 2^r$, which is a contradiction to the unique factorization of integers.

A.2 Since $(a, b) = 1$, it follows that $(a + b, ab) = 1$.

Therefore, $(a + b)^{\phi(ab)} \equiv 1 \pmod{ab}$.

Expanding the left hand side, $a^{\phi(ab)} + b^{\phi(ab)} \equiv 1 \pmod{ab}$.

A.3 We consider the fractions

$$\frac{1}{n}, \frac{2}{n}, \ldots, \frac{n-1}{n}, \frac{n}{n}.$$

If we reduce the fractions to lowest terms, then only $\phi(n)$ of the fractions will have n in their denominators.

We observe that this is true for all divisors of n. Let d be a divisor of n and $n = dd'$. Among the given fractions, only those whose numerators are mutiples of d', will have divisors of d in their denominators after reduction to lowest terms. They are

$$\frac{d'}{n}, \frac{2d'}{n}, \ldots, \frac{dd'}{n}.$$

There are d of them and they can be rewritten as

$$\frac{1}{d}, \frac{2}{d}, \ldots, \frac{d}{d}.$$

Therefore, among the reduced fractions, there are precisely $\phi(d)$ fractions having d in the denominator.

Since there are n fractions, we have $n = \sum_{d|n} \phi(d)$.

A.4 Given a divisor d of n, if there exists an element a of G of order d, then G has a cyclic subgroup H of order d generated by a. Now, any element b of H will satisfy the equation $b^d = 1$. Since H has d elements, by the given condition, there is no element of order d of G outside H. Observe that the number of generators of H is $\phi(d)$. Hence, if for a divisor d of n, there is an element of order d in G, then there are exactly $\phi(d)$ of them.

Now, every element of G generates a cyclic subgroup of order d, such that $d|n$. Therefore the number of elements of G is

$$\sum_{\substack{d|n \\ G \text{ has an element of order } d}} \phi(d).$$

Since from Exercise A.3 we have $n = \sum_{d|n} \phi(d)$, if for some d there is no element of order d, then it follows that the number of elements of G is strictly less than n, which is a contradiction.

Therefore, for every divisor d of n, G has an element of order d. Now, n itself being a divisor of n, there is an element of order n in G. This proves that G is cyclic.

3.2 Let a be an element of maximum order in G. Let n be the order of a. We claim that the order of any element of G will divide n. If possible, suppose b is an element of G whose order m does not divide n. This implies that there is a prime p such that the maximum power of p dividing m is strictly greater than the maximum power of p dividing n. That is, there is a prime p such that $p^i || m$ and $p^j || n$ with $i > j$.

Now, b^{m/p^i} is of order p^i and a^{p^j} is of order n/p^j.

Since, $(n/p^j, p^i) = 1$, the element $b^{m/p^i} a^{p^j}$ is of order $(n/p^j)p^i > n$, which is a contradiction to the maximality of n. This proves the first part of the exercise.

Let F be a finite field. Let α be an element of $F^* = F \setminus \{0\}$ of maximum order, say n. Then by the first part of the exercise, $\beta^n = 1$ for every $\beta \in F^*$. Since the polynomial $X^n - 1$ has at most n roots, it follows that the order of F^* is at most n. But $1, \alpha, \cdots, \alpha^{n-1}$ are in F^*. Therefore F^* is generated by α.

3.4 The number of distinct solutions modulo p for the polynomial congruence $f(x) \equiv 0 \pmod{p}$ is the same as the number of solutions of the equation

$$(x - \overline{1})(x - \overline{2}) \cdots (x - \overline{(p-1)}) - x^{p-1} + \overline{1} = \overline{0} \tag{14.5}$$

in the field $\mathbf{F}_p = \mathbf{Z}/p\mathbf{Z}$, where $\overline{a}$ denotes the residue class of a modulo p.

By Exercise 1.5 (Euler-Fermat) it follows that all the non-zero elements of $\mathbf{Z}/p\mathbf{Z}$ are solutions of $x^{p-1} - \overline{1} = \overline{0}$. Again, all the non-zero elements of $\mathbf{Z}/p\mathbf{Z}$ are solutions of $(x - \overline{1})(x - \overline{2}) \cdots (x - \overline{(p-1)}) = \overline{0}$.

Therefore, all the non-zero elements of $\mathbf{Z}/p\mathbf{Z}$ are solutions of the equation (14.5). But the equation is of degree less than $p - 1$ since the term x^{p-1} cancels out. Therefore the polynomial on the left-hand side of (14.5) is the zero polynomial in $\mathbf{F}_p[x]$.

This proves the first part of the exercise.

For the second part, we have to look at the constant term of the polynomial $f(x)$. The constant term is $(p-1)!+1$. From the first part of the exercise, it is divisible by p. This is precisely the statement of Wilson's theorem.

4.1 We consider the element $\overline{2}$ in the ring $R = \mathbf{Z}/6\mathbf{Z}$. If $\overline{2}|\ \overline{a}\cdot\overline{b}$ in R, then it is easy to see that $\overline{2}|\overline{a}$ or $\overline{2}|\overline{b}$. But since $\overline{2} = \overline{2}.\ \overline{4}$ where neither $\overline{2}$ nor $\overline{4}$ is a unit (the units in R being $\overline{1}$ and $\overline{5}$), it follows that $\overline{2}$ is not an irreducible element.

4.2 Let R be an integral domain and p a prime in R. Let $p = ab$ where $a, b \in R$. Since p is a prime, $p|a$ or $p|b$. If $p|a$, then $p = ab = (rp)b$ for some $r \in R$. This implies $p(1-rb) = 0$. Since R is an integral domain and p is non-zero, it follows that $rb = 1$. In other words, b is a unit. Similarly, $p|b$ will imply that a is a unit. Hence p is irreducible.

4.4 Let R be a PID and α an irreducible element of R. We have to show that if α divides the product of two elements of R, then α must divide one or both of them.

Let $\alpha|ab$ where $a, b \in R$. Suppose $\alpha \not| a$. Every ideal of R is principal and therefore, in particular, so is the ideal generated by α and b. Let it be generated by β.

Now, $\alpha = \beta r$ for some $r \in R$. Since α is irreducible, either β or r is a unit. If β is a unit, then 1 belongs to the ideal generated by α and b. This means, $x\alpha + yb = 1$ for some $x, y \in R$. Therefore, $ax\alpha + ayb = a$. Since $\alpha|ab$, that will imply $\alpha|a$, which is contrary to our assumption. Therefore, r is a unit and hence $(\alpha) = (\beta)$. Now, $b \in (\beta) = (\alpha)$ $\implies \alpha|b$ and we are through.

5.2 We observe that $f(x-1) = 8x^3 - 24x^2 + 18x - 3$. Taking $p = 3$, by Eisenstein Criterion (Theorem 5.7), we get $f(x-1)$ is irreducible in $\mathbf{Q}[x]$ and therefore so is $f(x)$.

6.1 Let K be a finite extension of a field F. Let $\alpha \in K$. Now, the sequence of elements $1, \alpha, \alpha^2, \cdots$ can not be linearly independent over F. In fact, if $[K : F] = n$, then the collection $\{1, \alpha, \alpha^2, \cdots, \alpha^n\}$ of $(n+1)$ elements will be linearly dependent over F. If $a_0 + a_1\alpha + \cdots + a_n\alpha^n$, $a_i \in F$ is a nontrivial linear combination which vanishes, then α satisfies the non-zero polynomial $a_0 + a_1x + \cdots + a_nx^n$, $a_i \in F$ and hence is algebraic over F.

6.11 Consider the polynomials $f(x_1, x_2, \cdots, x_{2p-1})$ and $g(x_1, x_2, \cdots, x_{2p-1})$ in $\mathbf{F}_p[x_1, x_2, \cdots, x_{2p-1}]$ defined by

$$f(x_1, x_2, \cdots, x_{2p-1}) = \sum_{i=1}^{2p-1} x_i^{p-1}$$

and

$$g(x_1, x_2, \cdots, x_{2p-1}) = \sum_{i=1}^{2p-1} a_i x_i^{p-1}.$$

Now, the sum of the degrees of the polynomials f and g is $2p-2 < 2p-1$. Theoreforе, by Remark 6.5 (since $f(0, \cdots 0) = g(0, \cdots, 0) = 0$), there is $(\alpha_1, \alpha_2, \cdots, \alpha_n) \in \mathbf{F}_p^n$ with not all α_i's zero such that

$$\sum_{i=1}^{2p-1} \alpha_i^{p-1} = 0 \tag{14.6}$$

and

$$\sum_{i=1}^{2p-1} a_i\alpha_i^{p-1} = 0. \tag{14.7}$$

For any $\alpha \in \mathbf{F}_p$, $\alpha^{p-1} = 1$ if and only if $\alpha \neq 0$. Therefore, from (14.6), exactly p of the α_i's, say $\alpha_{i_1}, \cdots, \alpha_{i_p}$ are non-zero and therefore, from (14.7), $a_{i_1} + \cdots + a_{i_p} = 0$.

7.1

$$\begin{aligned}\left(\frac{45}{1009}\right) &= \left(\frac{3^2}{1009}\right)\left(\frac{5}{1009}\right)\\ &= \left(\frac{5}{1009}\right)\\ &= \left(\frac{1009}{5}\right)(-1)^{\frac{1009-1}{2}\cdot\frac{5-1}{2}}\\ &= \left(\frac{1009}{5}\right)\\ &= \left(\frac{9}{5}\right) = 1.\end{aligned}$$

Hence 45 is a square modulo the prime 1009.

7.2 Suppose there are integers x and y satisfying $y^2 = x^3 + 23$. Now, $y^2 \equiv 0$, or $1 \pmod 4$ and $23 = 3 \pmod 4$. Since $x^3 \equiv 0, 1$, or $3 \pmod 4$, the only possibility is $x \equiv 1 \pmod 4$.

We write the given equation as $y^2 + 4 = x^3 + 27$. But $x^3 + 27 = (x+3)(x^2 - 3x + 9)$ and $(x^2 - 3x + 9) \equiv -1 \pmod 4$. Therefore there is some prime $p \equiv -1 \pmod 4$ such that $p \mid (x^2 - 3x + 9)$ and hence $p \mid y^2 + 4$. In other words, $y^2 \equiv -4 \pmod p$, which means $\left(\frac{-4}{p}\right) = 1 \Rightarrow \left(\frac{-1}{p}\right) = 1$, which is a contradiction. Hence the given Diophantine equation does not have any solution.

7.3 Let $p_1, p_2, \cdots, p_r$ be a finite set of primes of the form $8n - 1$. We consider the integer $N = (4p_1p_2\cdots p_r)^2 - 2$. If p is an odd prime divisor of N, then $\left(\frac{2}{p}\right) = 1$.

Therefore, by Theorem 7.1 (ii), $p = \pm 1 \pmod 8$. Writing $N = 2M$, we see that M is of the form $8n - 1$. Therefore, all the odd prime divisors of N can not be of the form $8n + 1$. Hence, N must have an odd prime divisor q of the form $8n - 1$. Since, q can not be among the primes $p_1, p_2, \cdots, p_r$, we are through.

B.1 Let σ be the size function which makes R an Euclidean domain. Let u be a non-zero non-unit in R such that $\sigma(u)$ is minimal. That is, for any other non-zero non-unit u', $\sigma(u) \leq \sigma(u')$. Given $\alpha \in R$, we can write $\alpha = qu + r$ with $q, r \in R$ and $\sigma(r) < \sigma(u)$.

The minimality of $\sigma(u)$ shows that r can not be among the non-zero non-units in R.

B.2 If possible, let R be a Euclidean domain. By Exercise B.1, there is a non-zero non-unit u in R such that for any $\alpha \in R$, $\exists r \in R^* \cup \{0\}$ with $u|(\alpha - r)$.

Considering (absolute value)^2 of elements of R, it can be easily seen that the unit group of R is $\{\pm 1\}$. Therefore, given any $\alpha \in R$, u divides one of the three elements α, $\alpha - 1$ or $\alpha + 1$. Since u is a non-unit, if we take $\alpha = 2$, we have $u|2$ or $u|3$. In other words, there is $v \in R$ satisfying $uv = 2$ or $uv = 3$.

We claim that $u \in \mathbf{Z}$. For if $u \notin \mathbf{Z}$, then $v \notin \mathbf{Z}$. Now, for any $r \in R$ with $r \notin \mathbf{Z}$, the square of the absolute value of r will be greater than three. Now this being the case with u and v, the square of absolute value of uv will be greater than 9 which contradicts the fact that $uv = 2$ or $uv = 3$. This establishes the claim.

Hence $u = \pm 2$ or $u = \pm 3$.

Now, we take $\alpha = \frac{1+\sqrt{-13}}{2}$, so that square of absolute values of none of the numbers α, $\alpha - 1$ and $\alpha + 1$ is divisible by 4 or 9. Hence u does not divide any of the three elements α, $\alpha - 1$ and $\alpha + 1$, which is a contradiction.

B.6 Let α be a generator of the cyclic group $\mathbf{F}_q^*$. Since $\mathbf{F}_q = \mathbf{F}_p(\alpha)$, the minimal polynomial of α over $\mathbf{F}_p$ is of degree $r = [\mathbf{F}_q : \mathbf{F}_p]$. (We observe that incidentally we have proved the fact that over $\mathbf{F}_p$, and similarly over any finite field, there exists an irreducible polynomial of any given degree.)

Now, any element of $\mathrm{Gal}(\mathbf{F}_q : \mathbf{F}_p)$ is determined by its value at α. But for any $\sigma \in \mathrm{Gal}(\mathbf{F}_q : \mathbf{F}_p)$, $\sigma(\alpha)$ has to be a conjugate of α. Hence $\mathrm{Gal}(\mathbf{F}_q : \mathbf{F}_p)$ can not have more than r elements. But the cyclic group generated by the Frobenius automorphism is of order r.

This is because $\alpha^{p^r} = \alpha$ and for any $s < r$, $\alpha^{p^s} = \alpha$ would imply the existence of p^r roots of the equation $x^{p^s} = x$ in F_q, which is not possible.

B.7 Two distinct polynomials of degree strictly less than q on $\mathbf{F}_q$ will not represent the same function. For, otherwise their difference, which will be a non-zero polynomial of degree less than q will have q roots in $\mathbf{F}_q$, which is not possible. But, the number of distinct polynomials in $\mathbf{F}_q[x]$ of degree less than q is q^q. Again, q^q is the number of functions $\mathbf{F}_q \to \mathbf{F}_q$. This proves the first part of the exercise.

Now, let R be a ring such that every function $f : R \to R$ is given by a polynomial in $R[x]$.

First, we observe that R must be finite. For, even if R is countably infinite, the set of functions from R to itself is not countable, whereas the set of polynomials on R is countable. In general, denoting the cardinality of R by $|R|$, the set of functions on R has cardinality $|R|^{|R|}$ and the second set has cardinality $\aleph_0|R|$ and these two cardinalities can be equal only when R is finite.

We claim that for any non-zero $r \in R$, the map $f_r : R \to R$ given by $x \mapsto rx$, is surjective.

Let $y \in R$ be any element. Consider the map

$$g_{r,y}(x) = \begin{cases} y, & \text{if } x = r \\ 0, & \text{if } x \neq r. \end{cases}$$

By our assumption, there is a polynomial $a_n x^n + \cdots + a_1 x + a_0 \in R[x]$ such that $g_{r,y}$ is given by this polynomial.

In other words,

$$a_n x^n + \cdots + a_1 x + a_0 = 0 \quad \text{for } x \neq r \tag{14.8}$$

and

$$a_n r^n + \cdots + a_1 r + a_0 = y \tag{14.9}$$

Putting $x = 0$ in (14.8), we have $a_0 = 0$. Therefore, from (14.9), we have $r(a_n r^{n-1} + \cdots + a_1) = y$. This establishes our claim. Since R is finite, this implies that f_r is injective. That means $rx = 0$ if and only if $x = 0$. Therefore, R is a finite integral domain and hence a field (by Exercise 0.2).

B.8 Considering the given equation with coefficients in $\mathbf{F}_7$, it is equivalent to the question whether $\overline{2}x^2 + \overline{3}x + \overline{1} = \overline{0}$ has solutions in $\mathbf{F}_7$. Multiplying both sides by $\overline{4}$ (which is the inverse of $\overline{2}$ in $\mathbf{F}_7$) and completing the square, it becomes $(x+\overline{6})^2 = \overline{4}$ and since $\left(\frac{4}{7}\right) = 1$, we conclude that there are integers x, y satisfying the given equation.

B.9 If possible, let the given Diophantine equation have a solution (x, y).

If possible, let $x \equiv 0 \pmod 2$. This will imply that $x^3 \equiv 0 \pmod 8$. Therefore, $y^2 \equiv 5 \pmod 8$. This is impossible, as it is not difficult to observe that square of any integer has to be $0, 1$ or $4 \pmod 8$.

Therefore, x must be odd.

If $x \equiv 1 \pmod 4$, then $y^2 \equiv 2 \pmod 4$, which again is impossible.

Therefore x has to be either 3 or 7 $\pmod 8$.

Let $x \equiv 3 \pmod 8$. We can write $y^2 = x^3 + 45$ as $y^2 - 2.6^2 = x^3 - 27 = (x-3)(x^2+3x+9)$. Now, we observe that $(x^2 + 3x + 9)$ is $3 \pmod 8$ and hence has a prime divisor $p \equiv \pm 3 \pmod 8$.

This prime p divides $y^2 - 2.6^2$
$\Longrightarrow \left(\frac{2.6^2}{p}\right) = 1 \Longrightarrow \left(\frac{2}{p}\right) = 1$, which is not possible as $p \equiv \pm 3 \pmod 8$.

The last possibility is that $x \equiv 7 \pmod 8$. Now, $y^2 - 2.3^2 = x^3 + 27 = (x+3)(x^2 - 3x + 9)$. In this case, $(x^2 - 3x + 9)$ is $-3 \pmod 8$ and hence has a prime divisor $p \equiv \pm 3 \pmod 8$.

As in the previous case, we derive $\left(\frac{2}{p}\right) = 1$, which is not possible as $p \equiv \pm 3 \pmod 8$.

Hence, we conclude that the given Diophantine equation does not have any solution.

8.1 We consider $\mathbf{Z}$ as a module over itself. For any positive integer n, the singleton set $\{n\}$ is of course linearly independent. But, if $n > 1$, then $\{n\}$ is neither a basis nor can it be extended to a basis.

Again, if p and q are two distinct primes in $\mathbf{Z}$, then there are integers a, b such that $1 = ap + bq$ and hence we see that the set $\{p, q\}$ generates $\mathbf{Z}$ as a module over itself. But p and q are not linearly independent. Therefore, $\{p, q\}$ does not contain any basis.

8.2 (a) Suppose M has a basis S. If S is empty, then by convention, M is the free module $\{0\}$. So assume S to be non-empty. For $s \in S$, it is easy to see that Rs is isomorphic to R as an R-module and M is direct sum of the collection of submodules $\{Rs\}_{s \in S}$.

Conversely, suppose $M = \bigoplus_{s \in S} R$. For each element $s \in S$, let α_s be the element $\{r_i\}$ of $\bigoplus_{s \in S} R$ such that $r_i = 0$ if $i \neq s$ and $r_s = 1$. It is then easy to see that $\{\alpha_s : s \in S\}$ is a basis of M.

(b) Let $\mathtt{A}$ be a maximal ideal of R. Now, M/AM is a vector space over the field $F = R/A$. If S is an R- basis for M, then $\{s + AM : s \in S\}$ is an F- basis for M/AM.

8.3 We prove the result for one variable. The general result will trivially follow from that by induction.

Let R be Noetherian and $\mathtt{A}$ be an ideal of $R[x]$. Clearly the collection of the leading coefficients of all the polynomials in $\mathtt{A}$ will be an ideal in R. We call that ideal I. Since R is Noetherian, I is generated by a finite set, say $\{r_1, r_2, \cdots, r_m\}$. For each i, $1 \leq i \leq m$, we take one polynomial, say $f_i(x)$, from $\mathtt{A}$ with leading coefficient r_i. If d is the maximum among the degrees of the f_i's, then it is not difficult to see that $\mathtt{A} = \mathtt{B} \cap \mathtt{A} + \mathtt{A}'$ where $\mathtt{B}$ is the R-module generated by $1, x, \cdots, x^{d-1}$ and $\mathtt{A}' \subset \mathtt{A}$ is the ideal generated by $\{f_1(x), \cdots, f_m(x)\}$. Now, $\mathtt{B}$ being a Noetherian R-module, the submodule $\mathtt{B} \cap \mathtt{A}$ is finitely generated and hence so is $\mathtt{A}$.

8.4 If possible, suppose that R is not Noetherian. Then if $\mathcal{S}$ denotes the collection of all ideals of R which are not finitely generated, we have

$\mathcal{S} \neq \emptyset$. Applying Zorn's lemma we get a maximal element of T of $\mathcal{S}$. T can not be a prime ideal. Therefore, there are $a, b \in R$ such that $ab \in T$ but neither a nor b is in T. Now, $T + bR > T$ and hence by maximality of T, $T + bR$ is finitely generated. Let $T + bR$ be generated by $\{b_1, b_2, \cdots, b_s\}$. Let $b_i = t_i + a_i b$ where $t_i \in T, a_i \in R$ for $1 \leq i \leq s$.

Now consider the ideal $(T : bR) = \{r \in R : rb \in T\}$. Clearly, $T \subset (T : bR)$. Also $a \in (T : bR)$. Hence $T < (T : bR)$. Again, by maximality of T, $(T : bR)$ is finitely generated. Let the set $\{c_1, c_2, \cdots, c_l\}$ generate $(T : bR)$. Clearly, $d_i \stackrel{\text{def}}{=} c_i b \in T$ and T is generated by the set $\{t_1, t_2 \cdots, t_s, d_1, d_2, \cdots d_l\}$, contradicting the fact that $T \in \mathcal{S}$. Hence $\mathcal{S} = \emptyset$.

8.6 Let V^* denote the dual of V consisting of all the linear forms on V. It is easy to see that V^* is again a vector space of dimension n over K. In fact, given a basis $v_1, \cdots, v_n$ of V, the elements f_j, $j = 1, \cdots, n$ of V^* where $f_j(v_i) = \delta_{ij}$ form a basis (called the dual of the basis of $v_1, \cdots, v_n$) of V^*. Consider the map $\phi : V \to V^*$ where for $b \in V$, $\phi(b)$ in V^* is the homomorphism given by $x \mapsto B(x, b)$. Clearly, ϕ is a homomorphism. Since B is non-degenerate, ϕ is injective. By dimension consideration, we see that ϕ must be surjective. Given a basis $\omega_1, \cdots, \omega_n$ of V, let f_j be the corresponding dual basis as described above. For each j, there is $\omega'_j \in V$ such that $\phi(\omega'_j) = f_j$. Now, $B(\omega_i, \omega'_j) = f_j(\omega_i) = \delta_{ij}$.

10.2 Let $n = 4k + 1$. Now, $\alpha = \frac{1+\sqrt{n}}{2}$ is an element in the quotient field of the given ring.

We have

$$\begin{aligned} \alpha^2 &= \frac{1 + 2\sqrt{n} + n}{4} \\ &= \frac{2\sqrt{n} + 4k + 2}{4} \\ &= k + \alpha. \end{aligned}$$

Thus, $\alpha^2 - \alpha - k = 0$. So we have found an element α in the quotient field of the given ring $R = \mathbf{Z} + \mathbf{Z}(\sqrt{n})$ such that α is integral over R and $\alpha \notin R$. Hence by Exercise 10.1, R is not a UFD.

C.1 Let π be a Gaussian prime which divides $a+ib$ in the ring of Gaussian integers $\mathbf{Z}[i]$. Now, π divides $(a+ib)(a-ib) = a^2+b^2 = c^2$ an even number of times. If π divides both $a+ib$ and $a-ib$ in $\mathbf{Z}[i]$, then π divides $2a$ in $\mathbf{Z}[i]$. Since a, b and c do not have any common factor, c has to be odd and $1 = \gcd(2a, c) = m(2a) + nc$, for some integers m and n. Therefore, π divides $m(2a) + nc = 1$ in $\mathbf{Z}[i]$, which is a contradiction to the fact that π is a Gaussian prime.

Therefore, π does not divide $a-ib$. Since, π divides $(a+ib)(a-ib) = a^2+b^2 = c^2$ an even number of times, we get our result.

C.2 Let $x^2+y^2 = z^2$, with the integers x, y, z pairwise coprime. (Once we find all such solutions, called the primitive solutions, other solutions will be of the form tx, ty, tz for $t \in \mathbf{Z}$.) We have $(x+iy)(x-iy) = z^2$ in $\mathbf{Z}[i]$. Since $\mathbf{Z}[i]$ is a UFD, by Exercise C.1, $x+iy$ is of the form $u(c+id)^2$ where u is a unit in $\mathbf{Z}[i]$ and c, d are integers. Since the only units in $\mathbf{Z}[i]$ are $\pm1, \pm i$, we have

$$x = \pm(c^2-d^2),\ y = \pm 2cd,\ z = \pm(c^2+d^2). \tag{14.10}$$

Conditions on x, y, z forces that c, d are relatively prime and not both of c and d are odd. Conversely, with such choice of c, d one gets primitive solutions of our equation $x^2+y^2 = z^2$ given by (14.10).

C.4 (a) Since, $N(a+b\omega) = a^2 - ab + b^2 = (a+b\omega)\overline{(a+b\omega)}$ we see that for two elements $\delta, \theta \in \mathbf{Z}[\omega]$, $N(\delta\theta) = N(\delta)N(\theta)$.

Given $\alpha, \beta \in \mathbf{Z}[\omega]$ such that $\alpha \neq 0$, observing that $\alpha \cdot \overline{\alpha} \in \mathbf{Z}$, we have $\beta/\alpha = (\beta\overline{\alpha})/(\alpha\overline{\alpha}) = r_1 + r_2\omega$, where r_i's are in $\mathbf{Q}$.

There are integers m_1 and m_2 such that $|r_i - m_i| \leq \frac{1}{2}$ for $i = 1, 2$. Now, writing $\kappa = m_1 + m_2\omega$, we have

$$N(\frac{\beta}{\alpha} - \kappa) = (r_1-m_1)^2 - (r_1-m_1)(r_2-m_2) + (r_2-m_2)^2 < 1.$$

Therefore, writing $\rho = \beta - \kappa\alpha$, if $\rho \neq 0$, then

$$N(\rho) = N(\alpha \cdot (\beta/\alpha - \kappa)) = N(\alpha) \cdot N(\beta/\alpha - \kappa) < N(\alpha).$$

(d) We do the second part. Let q be a rational prime such that $q \equiv 1 \pmod 3$. We have

$$\left(\frac{-3}{q}\right) = (-1)^{\frac{q-1}{2}} \left(\frac{q}{3}\right) (-1)^{\frac{q-1}{2}\frac{3-1}{2}} = 1.$$

Therefore, there is $n \in \mathbf{Z}$ such that $n^2 = -3 \pmod q$. This means q divides $(n+\sqrt{-3})(n-\sqrt{-3}) = (n+1+2\omega)(n-1-2\omega)$. Since $\frac{2}{q} \notin \mathbf{Z}$, it follows that q divides neither of the factors $(n+1+2\omega)$ and $(n-1-2\omega)$. Therefore q is not a prime in $\mathbf{Z}[\omega]$. Let α be a non-unit which divides q in $\mathbf{Z}[\omega]$, then $N(\alpha)$ divides q^2 in $\mathbf{Z}$. If $N(\alpha) = q^2$, then α is an associate of q. Therefore, if $q = \alpha\alpha'$, where neither α nor α' is a unit, then the only possibility is $q = N(\alpha) = \alpha\bar{\alpha}$.

(e) Putting $x = 1$ in the identity $x^2 + x + 1 = (x - \omega)(x - \omega^2)$, we have $3 = -\omega^2(1-\omega)^2$. This implies $9 = N((1-\omega)^2)$ and hence $(1-\omega)$ is a prime.

11.1 It is easy to see that A' is an R-module. Since A is a fractional ideal, A' is non-zero. Also, there is a non-zero element r of R such that $rA \subset R$. Therefore, if we choose a non-zero element x of A, then rx is a non-zero element of R. Since A is an R-module, $rx \in A$ as well and hence for any $y \in A'$, $rxy \in R$. Thus we have found a non-zero element rx of R such that $rxA' \subset R$. Hence A' is a fractional ideal.

11.3 We observe that $P^n + aR$ being an ideal containing P^n and not contained in P must be equal to R.

11.5 Let $\wp_1, \wp_2, \cdots, \wp_n$ be all the prime ideals (distinct) of a Dedekind domain R.

Choose $x_j \in \wp_j \setminus \wp_j^2$ for $1 \leq j \leq n$. By Chinese remainder theorem, for every j, $1 \leq j \leq n$, there is $y_j \in R$ such that

$$\begin{aligned} y_j &\equiv x_j \pmod{\wp_j^2} \\ y_j &\equiv 1 \pmod{\wp_i} \text{ for } i \neq j. \end{aligned}$$

Then, for every j, $1 \leq j \leq n$, we have $y_j R = \wp_j$. Since all the prime ideals are principal, so is any ideal and it follows that R is a PID.

11.6 It is known (Theorem 4.3) that a PID is a UFD. For a Dedekind domain R this also follows from the fact that in the ideal level there is unique factorization.

Therefore, if possible, suppose that R is a Dedekind domain which is a UFD but not a PID.

Now, all the prime ideals of R can not be principal. Let P be a non-principal prime ideal of R.

Let

$$\mathcal{S} = \{I : I \text{ is a non-zero ideal of } R \text{ such that } PI \text{ is principal}\}.$$

Let $\alpha\ (\neq 0) \in P$. Then $\alpha R \subset P$ and hence there is a non-zero ideal I_1 of R such that $\alpha R = PI_1$.

Therefore, $\mathcal{S}$ is non-empty and hence S has a maximal element M, say. Let $PM = \beta R$.

We claim that β is irreducible. If $\beta = \gamma\delta$, then $\beta R = (\gamma R)(\delta R)$. Therefore, P divides γR or δR. If P divides γR, then $\gamma R = PJ$ for some ideal J dividing M. Maximality of M implies that $M = J$. Therefore, $\delta R = R$, which implies that δ is a unit. Similarly if P divides δR, then γ is a unit. This establishes the claim.

Choose $\theta_1 \in P \setminus \beta R$ and $\theta_2 \in M \setminus \beta R$.

Then, $\theta_1\theta_2 \in \beta R$ but $\theta_1 \notin \beta R$ and $\theta_2 \notin \beta R$ which means

$$\beta | \theta_1\theta_2, \text{ but } \beta \nmid \theta_1, \beta \nmid \theta_2,$$

which is a contradiction, since R being a UFD, the irreducible element β has to be a prime.

12.1 It is sufficient to show that $N(\mathtt{A})N(\mathtt{P}) = N(\mathtt{A})N(\mathtt{P})$ where $\mathtt{P}$ is a maximal ideal of $\mathbf{O}_K$.

For a finite set S, denoting the number of elements of S by $|S|$, we have

$$|\mathbf{O}_K/\mathtt{AP}| = |\mathbf{O}_K/\mathtt{A}| \cdot |\mathtt{A}/\mathtt{AP}|. \tag{14.11}$$

Now, $\mathtt{A}/\mathtt{AP}$ is an $\mathbf{O}_K$-module annihilated by $\mathtt{P}$. Therefore, it is a vector space over $\mathbf{O}_K/\mathtt{P}$. A proper subspace of this vector space is of the form $\mathtt{A}'/\mathtt{AP}$ where $\mathtt{A}'$ is an ideal of $\mathbf{O}_K$ such that $\mathtt{AP} < \mathtt{A}' < \mathtt{A}$.

From Remark 11.5, $\mathtt{A}' = \mathtt{AB}$ for some ideal $\mathtt{B} \neq \mathbf{O}_K$ of $\mathbf{O}_K$. Since, $\mathtt{AP} < \mathtt{A}'$, there is no such ideal $\mathtt{B}$. Hence, $\mathtt{A}/\mathtt{AP}$ is a vector space of dimension one over $\mathbf{O}_K/\mathtt{P}$ and this implies that $|\mathtt{A}/\mathtt{AP}| = |\mathbf{O}_K/\mathtt{P}|$. Therefore, from (14.11), $|\mathbf{O}_K/\mathtt{AP}| = |\mathbf{O}_K/\mathtt{A}| \cdot |\mathbf{O}_K/\mathtt{P}|$.

12.2 Let $\mathtt{A}$ be an ideal of $\mathbf{O}_K$ such that $N(\mathtt{A}) \leq m$. Let $\mathtt{A} = \mathtt{P_1}^{r_1}\mathtt{P_2}^{r_2}\cdots\mathtt{P_1}^{r_l}$ be the decomposition of $\mathtt{A}$ into products of prime ideals of $\mathbf{O}_K$. Since $N(\mathtt{P_i}) \geq 2$, by Exercise 12.1, the condition $N(\mathtt{A}) \leq m$ will imply that the r_i's are bounded above. Now, we observe that the prime ideal P_i appearing in the factorization of $\mathtt{A}$ will contain a rational prime $p_i < m$. This is so because the norm of a prime ideal containing a rational prime p is p^d for some $1 \leq d \leq n$ where n is the degree of K. Again, there are no more than n prime ideals of $\mathbf{O}_K$ containing a rational prime p. Thus in the prime factorization of an ideal $\mathtt{A}$ of $\mathbf{O}_K$, only finitely many prime ideals can appear with uniformly bounded exponent if the condition $N(\mathtt{A}) \leq m$ is imposed. Hence the result.

12.3 A vector v belonging to the lattice H is of the form

$$r_1v_1 + r_2v_2 + r_3v_3 + r_4v_4$$

$$= (r_1m + r_3a + r_4b, r_2m + r_3b - r_4a, r_3, r_4),$$

where r_i's are in $\mathbf{Z}$.

Since $a^2 + b^2 + 1 \equiv 0 \pmod{m}$, it is easy to observe that the integer $|v|^2$ is a multiple of m.

Now $\mathcal{B}$ is a convex body symmetric about the origin and the volume of $\mathcal{B}$ is $2\pi^2 m^2 > 2^4 m^2 = 2^4 \nu(H)$. Therefore, by Corollary 12.8.1, $\mathcal{B}$ contains a non-zero lattice point $u = (l_1, l_2, l_3, l_4)$, say.

Now, $0 < |u|^2 < 2m$ and $|u|^2$ is a multiple of m forces that $|u|^2 = m$, that is, $l_1^2 + l_2^2 + l_3^2 + l_4^2 = m$.

Finally, since $l_1^2 + l_2^2 + l_3^2 + l_4^2 = m$ implies that

$$2m = (l_1 + l_2)^2 + (l_1 - l_2)^2 + (l_3 + l_4)^2 + (l_3 - l_4)^2,$$

we are through.

12.5 Here, with the notations as in Theorem 12.10, we have $r_2 = 1$ and $n = 2$. From Remark 13.2, $|d(K)| = 20$. Hence,

$$(\frac{4}{\pi})^{r_2} \frac{n!}{n^n} \mid d(K) \mid^{1/2} = \frac{4\sqrt{5}}{\pi} < 3.$$

Therefore, it is enough to find the prime divisors of $2\mathbf{O}_K$. Now, $2\mathbf{O}_K = (2, 1 + \sqrt{-5})^2$. By Remark 9.2, the ideal $(2, 1 + \sqrt{-5})$ is not principal. It can also be seen directly as follows. If $(2, 1 + \sqrt{-5})$ were principal, say (α), then writing $\alpha = a + b\sqrt{-5}$ we would have $\mid N_K(\alpha) \mid = N((\alpha)) = 2 \Rightarrow a^2 + 5b^2 = \pm 2$, which is impossible. Hence, the class number of $\mathbf{Q}(\sqrt{-5})$ is 2.

D.1 Let $\mathtt{A} = \wp_1^{m_1} \wp_2^{m_2} \cdots \wp_r^{m_r}$ and $\mathtt{B} = \wp_1^{n_1} \wp_2^{n_2} \cdots \wp_r^{n_r}$ where $\wp_i$'s are prime ideals and m_i, n_i are non-negative integers.

For all i, $1 \leq i \leq r$, choose

$$\alpha_i \in (\wp_1^{m_1+1} \cdots \wp_{i-1}^{m_{i-1}+1} \wp_i^{m_i} \wp_{i+1}^{m_{i+1}+1} \cdots \wp_r^{m_r+1})$$

such that

$$\alpha_i \notin (\wp_1^{m_1+1} \cdots \wp_{i-1}^{m_{i-1}+1} \wp_i^{m_i+1} \wp_{i+1}^{m_{i+1}+1} \cdots \wp_r^{m_r+1}).$$

If we take $\omega = \alpha_1 + \cdots + \alpha_r$, we get that the gcd $(\mathtt{AB}, (\omega))$ is $\mathtt{A}$.

D.2 Suppose an ideal $\mathtt{A}$ of $\mathbf{O}_K$ is given. Now, $\mathtt{A}^{-1}$ is a fractional ideal of $\mathbf{O}_K$ and hence there is α $(\neq 0) \in \mathbf{O}_K$ such that $\alpha\mathtt{A}^{-1} \subset \mathbf{O}_K$. Writing $\alpha\mathtt{A}^{-1} = \mathtt{B}$, we have $AB = \alpha\mathbf{O}_K$. Therefore, by Exercise D.1, there exists $\omega \in \mathbf{O}_K$ such that the greatest common divisor $(\mathtt{AB}, (\omega))$ is $\mathtt{A}$. In other words (see Remark 11.5), $\mathtt{A} = \mathtt{AB} + \omega\mathbf{O}_K = \alpha\mathbf{O}_K + \omega\mathbf{O}_K$.

D.3 The minimal polynomial of ζ_p over $\mathbf{Q}$ is $x^{p-1} + x^{p-2} + \cdots + 1$.

Therefore, $Tr(\zeta_p^j) = Tr_{K/\mathbf{Q}}(\zeta_p^j) = -1$ for $j = 1, \cdots, p-1$.

Again, $Tr(1) = p - 1$.

Therefore,

$$Tr(1-\zeta_p) = Tr(1-\zeta_p^2) = \cdots = Tr(1-\zeta_p^{p-1}) = p. \qquad (14.12)$$

Putting $x = y + 1$,

$$\frac{x^p - 1}{x - 1} = \frac{(y+1)^p - 1}{y} = y^{p-1} + py^{p-2} + \cdots + p \qquad (14.13)$$

and hence

$$\begin{aligned} N(\zeta_p - 1) &= (-1)^{p-1}p \\ \Longrightarrow N(1-\zeta_p) &= p. \end{aligned}$$

Thus,

$$(1-\zeta_p)(1-\zeta_p^2)\cdots(1-\zeta_p^{p-1}) = p. \qquad (14.14)$$

Now, (14.14) implies that p belongs to the principal ideal $(1-\zeta_p)\mathbf{O}_K$ and hence

$$p\mathbf{Z} \subset (1-\zeta_p)\mathbf{O}_K \cap \mathbf{Z}.$$

If $p\mathbf{Z} \neq (1-\zeta_p)\mathbf{O}_K \cap \mathbf{Z}$, then $p\mathbf{Z}$ being a maximal ideal of $\mathbf{Z}$, we have

$$\mathbf{Z} = (1-\zeta_p)\mathbf{O}_K \cap \mathbf{Z}.$$

$$\Longrightarrow 1 \in (1-\zeta_p)\mathbf{O}_K.$$

$$\Longrightarrow 1-\zeta_p \text{ is a unit of } \mathbf{O}_K.$$

$$\Longrightarrow N(1-\zeta_p) = \pm 1.$$

This is a contradiction to 14.14.

Therefore,

$$p\mathbf{Z} = (1-\zeta_p)\mathbf{O}_K \cap \mathbf{Z} \tag{14.15}$$

This proves Part (i).

Now, let y be any element of $\mathbf{O}_K$.

A conjugate of $y(1-\zeta_p)$ is of the form $y_j(1-\zeta_p^j)$ for $1 \le j \le p-1$, where y_j is a conjugate of y. Since $1-\zeta_p^j = (1-\zeta_p)(1+\zeta_p+\cdots+\zeta_p^{j-1})$, it follows that

$$Tr(y(1-\zeta_p)) \in p\mathbf{Z}. \tag{14.16}$$

Now we proceed to prove Part (iii).

We have

$$\alpha(1-\zeta_p) = a_0(1-\zeta_p) + a_1(\zeta_p - \zeta_p^2) + \cdots + a_{p-2}(\zeta_p^{p-2} - \zeta_p^{p-1}).$$

Therefore, $Tr(\alpha(1-\zeta_p)) = a_0 Tr(1-\zeta_p) = a_0 p$, by (14.12).

Hence, by Part (ii), that is, by (14.16) above, we have $a_0 p \in p\mathbf{Z}$ which in turn implies that $a_0 \in \mathbf{Z}$.

Now, $a_1\zeta_p + \cdots + a_{p-2}\zeta_p^{p-2} \in \mathbf{O}_K$. Multiplying by $\zeta_p^{p-1} = \zeta_p^{-1}$, we get $a_1 + \cdots + a_{p-2}\zeta_p^{p-3} \in \mathbf{O}_K$.

Proceeding as before, $a_1 \in \mathbf{Z}$. By repeated application of this argument we get $a_i \in \mathbf{Z}$ for all i.

D.4 We solve only part (iv). Suppose $p|f(a)$. Then by Part (iii), $a^m \equiv 1 \pmod p$ and $a^r \not\equiv 1 \pmod p$ for any proper divisor of m. Considering a as an element of $\mathbf{F}_p^*$, we get that $m|(p-1)$.

Conversely, if $p \equiv 1 \pmod m$, since $\mathbf{F}_p^*$ is cyclic of order $p-1$, there is an integer a whose multiplicative order mod p is m. Therefore, by Part (iii), $p|f(a)$.

Now, suppose we are given a set (possibly empty) of finitely many primes $\mathcal{S} = \{p_1, \cdots, p_r\}$ contained in the arithmetic progression $1, 1+m, 1+2m, \cdots$.

Write $s = mp_1 \cdots p_r$. (We have $s = m$, if $\mathcal{S}$ is empty.)

Let t be any integer. Then $f(st) = \pm 1 \pmod{s}$. This implies that $f(st) = \pm 1 \pmod{p_i}$ for $i = 1, \cdots, r$ and $f(st) = \pm 1 \pmod{m}$. Making $t \to \infty$, we have $f(st) \to \infty$. Therefore, we can choose t such that $f(st) \neq \pm 1$.

Let P be a rational prime dividing $f(st)$. Then, by the the first part, $P \equiv 1 \pmod{m}$. We have already seen that no p_i divides $f(st)$. Therefore P is a prime other than the p_i's such that $P \equiv 1 \pmod{m}$.

13.1 K is norm-Euclidean if for given $a, b \in \mathbf{O}_K$, with b non-zero, there exists $q \in \mathbf{O}_K$ such that

$$| N_{K/\mathbf{Q}}(a - qb)| <| N_{K/\mathbf{Q}}(b)|.$$

Since norm is multiplicative, the above condition can be written as

$$| N_{K/\mathbf{Q}}(ab^{-1} - q)| < 1.$$

Appendix: Lucas-Lehmer test

Here we shall see an application of our knowledge of number fields to primality testing.

It is easy to see that for integers $n > 1, a > 1$, if $a^n - 1$ is a prime, then $a = 2$ and n is a prime. It was asserted by Mersenne that for primes $p \leq 257$, the integer $M_p \stackrel{\text{def}}{=} 2^p - 1$ is a prime for precisely 11 such p's, these 11 primes being listed by him. Later, the statement was found to contain several mistakes. For instance, the first mistake to be discovered was that M_{61} is a prime which does not appear in Mersenne's list.

The following remarkable theorem gives a necessary and sufficient condition for testing primality of M_p for an odd prime p. The test can be easily implemented on a computer. Our proof here is due to Rosen [Ro 1988] as simplified by Bruce [Br 1993].

Theorem A.1. (**Lucas-Lehmer**) For an odd prime p, $M_p = 2^p - 1$ is a prime if and only if M_p divides S_{p-1} where the numbers S_n are defined by $S_1 = 4, S_n = S_{n-1}^2 - 2$ for $n \geq 2$.

Before embarking on the proof of the theorem, we prove two lemmas.

Let

$$\tau = \frac{1+\sqrt{3}}{\sqrt{2}}, \quad \overline{\tau} = \frac{1-\sqrt{3}}{\sqrt{2}}, \quad \omega = \tau^2 = 2+\sqrt{3}, \ \overline{\omega} = \overline{\tau}^2 = 2-\sqrt{3}.$$

We note that

$$\tau\overline{\tau} = -1, \ \omega\overline{\omega} = 1. \qquad (*.1)$$

Lemma A.1. For all integers $m \geq 1$, $S_m = \omega^{2^{m-1}} + \overline{\omega}^{2^{m-1}}$.

Proof: For $m \geq 1$, write $T_m = \omega^{2^{m-1}} + \overline{\omega}^{2^{m-1}}$.

We have $T_1 = \omega + \overline{\omega} = 4 = S_1$.

Now, from the definition of T_n, $T_n = T_{n-1}^2 - 2$ for $n \geq 2$. Since $T_1 = S_1$, the lemma follows from the definition of S_n. □

Lemma A.2. If M_p is a prime, then $\tau^{M_p+1} \equiv -1 \pmod{M_p}$ in the ring of algebraic integers **O**.

Proof: During this proof, we set $M = M_p$.

Since $\sqrt{2}\tau = 1 + \sqrt{3}$, we get that the following congruence holds in **O**.

$$\tau^M 2^{\frac{M-1}{2}} \sqrt{2} \equiv 1 + 3^{\frac{M-1}{2}} \sqrt{3} \pmod{M}. \qquad (*.2)$$

Since $M \equiv -1 \pmod 8$ as well as $M \equiv 1 \pmod 3$, we have the following congruences in $\mathbf{Z} \subset \mathbf{O}$.

$$2^{\frac{M-1}{2}} \equiv \left(\frac{2}{M}\right) = 1 \pmod{M}. \qquad (*.3)$$

and

$$3^{\frac{M-1}{2}} \equiv \left(\frac{3}{M}\right) = -1 \pmod{M}. \qquad (*.4)$$

Now, from (∗.2), (∗.3) and (∗.4), we have

$$\tau^M \sqrt{2} \equiv 1 - \sqrt{3} \pmod{M}.$$

In other words,

$$\sqrt{2}\left(\tau^M - \overline{\tau}\right) = M\theta,$$

for some $\theta \in \mathbf{O}$.

Since $M = M_p = 2^p - 1$, multiplying the above by $2^{(2p-1)/2}$, we have

$$\tau^M \equiv \overline{\tau} \pmod{M}.$$

$$\Rightarrow \tau^{M+1} \equiv \tau\overline{\tau} \equiv -1 \pmod{M}.$$

□

Proof of Theorem A.1: Let M_p be a prime for an odd prime p. Then, by Lemma A.2, we have the following congruence in **O**

$$\tau^{2^p} + 1 \equiv 0 \pmod{M_p}$$

$$\Rightarrow \omega^{2^{p-1}} + 1 \equiv 0 \pmod{M_p}$$

$$\Rightarrow \omega^{2^{p-2}} + \overline{\omega}^{2^{p-2}} \equiv 0 \pmod{M_p}. \qquad \text{(by } (*.1))$$

Hence, by Lemma A.1, $S_{p-1} = M_p\delta$, where $\delta \in \mathbf{O}$. Since $\delta \in \mathbf{Q}$ as well, we have $\delta \in \mathbf{Z}$.

Conversely, suppose M_p divides S_{p-1} for an odd prime p.

If possible, let M_p be not a prime. Let q be a prime divisor of M_p such that $q^2 \leq M_p$.

Since, M_p divides S_{p-1}, in the ring of integers $\mathbf{O}_K$ of $K = \mathbf{Q}(\sqrt{3})$ we have

$$\omega^{2^{p-1}} + 1 \equiv 0 \pmod{q}.$$

This implies ω has order 2^p in the group $\left(\frac{\mathbf{O}_K}{q\mathbf{O}_K}\right)^*$.

Since the order of the group $\left(\frac{\mathbf{O}_K}{M_p\mathbf{O}_K}\right)^*$ does not exceed $q^2 - 1$, we have

$$2^p \leq q^2 - 1 \leq M_p - 1 = 2^p - 2,$$

which is absurd. Hence, M_p is a prime. □

Bibliography

AM 1969. M. F. Atiyah and I. G. Macdonald, *Introduction to Commutative Algebra*, Addison-Wesley.

Ar 1994. Michael Artin, *Algebra*, Prentice Hall.

Br 1993. J. W. Bruce, *A Really Trivial Proof of the Lucas-Lehmer Test*, Amer. Math. Monthly, Vol. 100, 370-371.

Cl 1994. D.A. Clark, *A quadratic field which is Euclidean but not norm - Euclidean*, Manuscripta Math., Vol. 83, 327-330.

DPS 1996. C. Ding, D. Pei, A. Salomaa, *Chinese Remainder Theorem – Applications in Computing, Coding, Cryptography*, World Scientific.

Du 1969. U. Dudley, *Elementary Number Theory*, W. H. Freeman and Company, San Francisco.

EM 1999. Jody Esmonde and M. Ram Murty, *Problems in Algebraic Number Theory*, Springer-Verlag.

GMM 1987. R. Gupta, M. Murty, V. Murty, *The Euclidean algorithm for S- integers*, Canad. Math. Soc. Conference Proc., Vol. 7, 189-201.

HW 1981. G.H. Hardy & E. M. Wright, *An introduction to the Theory of Numbers*, 5th edition, Oxford University Press.

He 1975. I. N. Herstein, *Topics in Algebra*, 2nd edition, Wiley, New York.

Hu 1982. L. K. Hua, *Introduction to Number Theory*, Springer-Verlag.

IR 1982. Kenneth Ireland and Michael Rosen, *A Classical Introduction to Modern Number Theory*, Springer-Verlag.

La 1993. Serge Lang, *Algebra*, 3rd edition, Addison-Wesley.

Le 1995. Franz Lemmermeyer, *The Euclidean algorithm in algebraic number field*, Expo. Math. Vol. 13, 385-416.

LN 1983. R. Lidl and H. Niederreiter, *Finite Fields*, Encyclopedia of mathematics and its applications, Vol. 20, Addison-Wesley.

Ma 1977. Daniel A. Marcus, *Number Fields*, Springer-Verlag.

Mu 1994. C. Musili, *Introduction to Rings and Modules*, Narosa.

Na 1990. W. Narkiewicz, *Elementary and Analytic Theory of Algebraic Numbers*, 2nd edition, Springer -Verlag.

Nat 1996. Melvyn B. Nathanson, *Additive Number Theory: Inverse problems and the geometry of sumsets*, Springer-Verlag.

NRRL 1966. Raghavan Narasimhan, S. Raghavan, S.S. Rangachari and Sundar Lal, *Algebraic Number Theory*, Tata Institute.

Ro 1988. M. I. Rosen, *A Proof of the Lucas-Lehmer Test*, Amer. Math. Monthly, Vol. 95, 855-856.

Sa 1967. P. Samuel, *Théorie Algébrique des Nombers*, Hermann & Cie.

Sa 1968. P. Samuel, *Unique Factorization*, Amer. Math. Monthly, Vol. 75, 945-952 .

Se 1973. J. -P. Serre, *A course in Arithmetic*, Springer-Verlag.

Sp 1994. Karlheinz Spindler, *Abstract Algebra with Applications, Vol. II*, Marcel Dekker.

Wa 1982. Lawrence C. Washington, *Introduction to Cyclotomic Fields*, Springer-Verlag.

We 1973. P. J. Weinberger, *On Euclidean rings of algebraic integers*, Proc. Symp. Pure Math., Vol. 24, 321-332.

Wy 1972. B.F. Wyman, *What is a reciprocity law?*, Amer. Math. Monthly, Vol. 79, 571-586.

Index